WALLPAPER

ITS HISTORY, DESIGN AND USE

Courtesy M. H. Birge and Sons Company

THE PEACOCK DESIGN, ONE OF THE MOST SUMPTUOUS MODERN AMERICAN PRODUCTIONS.

WALLPAPER

ITS HISTORY, DESIGN AND USE

BY

PHYLLIS ACKERMAN, PH.D.

WITH FRONTISPIECE IN COLOR AND NUMEROUS ILLUSTRATIONS FROM PHOTOGRAPHS

NEW YORK
FREDERICK A. STOKES COMPANY
MCMXXIII

Printed in the United States of America

PREFACE

An art that has become an industry often ceases to be regarded as an art, and usually with some justice, for competition, with the resulting struggle for cheap production and the violent fluctuations in fashion induced artificially for the sake of profit, are not conducive to the maintenance of standards. So wallpaper has been looked upon largely as a commodity and has been almost completely neglected by students of the decorative arts. Its glorious past, however, is not depressed by commercialism and sporadic efforts are still being made to return it to the level of its own tradition. The possibilites of a decorative art of distinction are there, as the French designers of the Eighteenth Century demonstrated. Only interest and understanding are needed for their fruition again.

Wallpaper is an important decorative art because, if for no other reason, it is so ubiquitous. Because of its general use it is one of the most important arts for general education in design. Bad wallpaper can do more than any other one decorative art to stultify taste; good, to stimulate it. This book is offered in the hope that it will contribute to the interest in and understanding of wallpaper, and so help to accelerate that revival of it as an art, which is already initiated at several points.

Mr. Walter Pach gave sympathetic help in assembling the material on modernist papers and I have had indispensable assistance throughout from Arthur Upham Pope.

CONTENTS

CHAPTER		PAGE
	Preface	xiii
	Introduction	xv
I	The History of Wallpaper	1
II	Wallpapers in Early American Homes	65
III	The Modernist Wallpaper	81
IV	The Manufacture of Wallpaper	94
V	The Problems of Wallpaper Designs	106
VI	The Importance of Wall Decoration	129
VII	Color Problems	143
VIII	The Importance of Texture in Wall Decoration	166
IX	Line and Scale in Wallpaper Pattern	179
X	Wallpapers for Period Rooms	192
XI	Wallpapers for Other Rooms	228

APPENDIX

Wallpaper Designers	241
Wallpaper Printers and Dealers	247
Bibliography	257
Index	265

ILLUSTRATIONS

The Peacock Design, One of the Most Sumptuous Modern American Productions (*in color*) . . . . *Frontispiece*

FACING PAGE

Fragment of an Early English Wallpaper, Sixteenth to Seventeenth Century 12

An English Wallpaper of the Period of Queen Elizabeth 13

Jean Pillement, who Made Designs for Le Sueur, is especially Famous for His Chinoiseries . . 20

A French Eighteenth Century Reproduction of a Chinese Bird and Flower Design 21

The Bird and Flower Wallpaper Designs of China were Derived from an Old Tradition in Chinese Painting 28

The Landscape Designs of the Chinese were Beginning to be Very Much in Demand in the Reign of Queen Anne 29

The Chinese Landscape Papers were Adapted from the Decorative Landscape Paintings of the Ming Period 36

Many of the Papers Showing Scenes of Daily Life are Very Similar to the Mirror Paintings that were in Vogue at the Same Time 37

One Panel from the Life of Psyche Designed by Laffitte and Printed by Dufour in the First Quarter of the Nineteenth Century . . . 44

FACING PAGE

The Irresponsible French Interpretations of Chinese Decoration were Repeated with Many Variations for Almost a Century. Early Nineteenth Century Examples 45

A Page from Rowlandson's Borders for Rooms and Screens 60

An Early Nineteenth Century French Paper Shows Scenes from the Bosporus. In an Old House at Marblehead, Mass. 68

The Paper Showing Scenes of Paris Brought Together All the Famous Buildings of the City. In an Old House at Salem, Mass. . . . 69

The Scenes of America Included Boston Harbor. A Modern Reprinting of an Early Nineteenth Century French Paper 76

Conventional Block Patterned Papers were Frequently Used in Early American Houses. A Modern Reprinting of an Old Design . . 77

A Modern German Design that has Felt the Influence of Cubism 84

A Modernist Paper from Germany 85

Making the Rolls from which Wallpaper is Printed 100

The Twelve-Color Printing Machine at Work . 101

Perspective may be Adapted to a Flat Surface by Conventionalization in Horizontal Bands . 108

Plant Motives, either in Their Naturalistic Drawing or Conventionalized into a Pattern, are a Fundamental Material of Wallpaper Design 109

Textile Patterns were the Earliest Used in Wall-

FACING PAGE

papers. Facsimile of Portion of Early Wallpaper Found in Master's Lodge, Christ College, Cambridge. Sixteenth Century . . . 124

Energetic Curves are Nowhere More Vitally Drawn than in Sixteenth Century Persian Rugs 125

The Chinese Papers are Entirely in Keeping with Chippendale even where only Part of the Furniture is in the Chinese Manner . . . 196

An Antique Hand-Painted Chinese Bird and Flower Paper is a Beautiful and Fitting Background for Sheraton 197

A Characteristic Louis XV French Wallpaper . 204

An Unusual Adaptation of Chinese Motives Makes a Characteristic Wall Decoration in a Louis XVI Room 205

The Empire is the Period of the Greatest Vogue of Landscape Paper. An Old Example . . 220

Gay and Amusing Designs that would be too Obtrusive Elsewhere can Find a Happy Place in the Kitchen 228

INTRODUCTION

WALLPAPER was a product of necessity. Some sort of covering for the walls of the early European houses was absolutely essential to make the rooms livable. The Greeks and the Romans of the Imperial period had been lavish in the use of elaborately woven and embroidered draperies of wool and linen in both their public buildings and their homes in spite of the fact that their walls were of beautifully polished marble or carefully finished stone or plaster. It was even more essential for the early Europeans, the Franks and Burgundians and the other successors to the shattered Roman Empire, to have some kind of wall hangings, because their climate was colder and their rougher, cruder walls more emphatically called for some mitigating covering. Woven materials, of course, were the obvious solution of the problem and so the magnificent tapestries of the

Gothic period were gradually developed. Rich silks, too, were imported from the East for the purpose and later were manufactured in Europe. Wood paneling, also, came to be utilized, especially oak paneling richly carved and painted. And all of these devices, tapestries and silks and panels, were successful in creating an atmosphere of comfort in those rather rigorous rooms.

But all of these took time and skill and expensive materials, and so their price was high and they were beyond the reach of all but the most wealthy. For the householders of more modest requirements, there were simpler woolen weaves, known in general as serges, but even these, because of the price of wool and the long labor of weaving them by hand, were costly. There was great need for some kind of wall covering that would make the walls look finished and warm and livable and still not have to cost a forbidding amount.

Paper fulfilled just these requirements. It gave an appearance of decorative finish and comfort to a room and still was relatively cheap. Once introduced, its vogue spread rapidly. Painted papers, block-printed papers, flock papers and finally roller-printed papers came into the market in ever increasing quantities. At first only a make-shift for the poorer classes, paper was soon seen to be so interesting and to have such decorative possibilities that it quite supplanted all other wall decorations. Through its own intrinsic merit it rose from its humble beginnings as the economical substitute for woven wall coverings and the more or less successful imitations of them to a level of artistic importance where it commanded the services of painters and engravers of note.

A few years ago there was a curious reaction against papered walls. It was partly an outcome of timidity. Shaken by the accepted indictment of our own bad taste we refused to venture into the perilous problems of pattern and of color. In evasion we extolled the beauty of plain walls and in defense cited the old wood paneling and, more especially, plain plaster. In calling upon this evidence we conveniently forgot that the panels of the Fifteenth Century were not unpatterned but were carved and gaily painted in gold and rich vermilion, and the panels of the Eighteenth Century at their best were usually pictorial. When we fell back on plain plaster we ignored the fact that the old plaster was hand wrought and quite remote from the bald mechanical surfaces with hard or gritty texture that we have recently had the temerity to expose. With the characteristic irrelevance of a weakness protecting itself our timidity of taste also fortified its defense of plain walls by an appeal to the Craftsman movement, in spite of the fact that William Morris, the master craftsman, urged and fostered design and himself made some of the finest patterns for wallpapers produced in a hundred years. Now, bored by this self-inflicted barrenness, we have turned again to the interest of pattern and of color and are redeeming our walls from the mechanical monotony of flat paint.

Wallpaper does contribute to decorative qualities that no other wall covering can give in quite the same way. It has, in some of its designs, a freshness and crispness that it shares in part with printed cottons, but which it can express even more directly than they. It gives a more personal character than a silk brocade, a more intimate and informal manner to the room, and yet it is seldom frivolous or trivial. It becomes a coherent part of the material of the room, one with the walls, as no woven fabric ever can, so that it seems solid, strong and structural, but it is never cold and hard as any but the most beautifully wrought plasters unavoidably are. It has besides a range of character from the gay and whimsical through the luxurious to the solemn and dignified that is not available in any other material.

This book is a consideration of the decorative qualities of wallpaper first as revealed in its historical development, second as limited by its present mechanical production, third as determined by the requirements of good design, and fourth as realized in its appropriate use.

The characteristic quality of any art is best revealed in its development. Each stage of the history of an art is the revelation of a new phase of its possibilities. When technique is still tentative the designer is held within limitations that keep the art true to its own character. Thus the tapestry weavers of the Fourteenth and Fifteenth Centuries restricted themselves to designs fitted to the simple technique of their art and escaped the later fallacy of trying to imitate painting. When technical control is first achieved the discovery of the new opportunities stimulates the artists to their utmost efforts. The early Renaissance masters, conscious of their newly acquired mastery of drawing and of pigments, painted with a brilliance and a power that was not equaled for many generations. The novelty of his problems in the new technique forces the artist, too, to be original. Decadence and its sterility appear only when mechanical ingenuity has overcome all obstacles and a long tradition has taken the place of fresh invention. So the history of wallpaper is interesting not only as the story of the rise of an art and an industry but as the progressive unfolding of the decorative possibilities of paper as a wall covering.

This does not mean, of course, that only old paper made before the perfection of modern machinery is decoratively good, or even that all old wallpaper is good at all. The chances are perhaps greater that an old hand-painted or wood-blocked paper will have artistic merit than that an ordinary machine product will be of distinctive worth; but decorative success does not depend primarily on the means of production but on the quality of the design. It is the beauty of the pattern that determines the real worth of any work of decorative art.

Given good wallpaper of good design, the final problem is how to use it in order to realize best its decorative possibilities. Every room has its specific needs. Size, height, light and character, each makes its own demand. The texture of the paper is an opportunity and a pitfall. The way it is put on the wall—straight, paneled, with dadoes or bordered—will contribute to its effect. If it is a period room historical correctness adds a factor. The successful use of wallpaper as a decoration is itself an art.

There is at present a revival of interest in wallpaper as a decoration, and especially in the old patterns of the late Eighteenth and early Nineteenth Centuries. However, if this revived interest is to stimulate the present art of wallpaper, so that it will universally attain again the level of its great period, it is necessary that there be again a thorough understanding of its character, its possibilities and its needs. This book hopes to contribute to that understanding and so to the encouragement of the production and use of really fine wallpaper.

WALLPAPER

CHAPTER I

THE HISTORY OF WALLPAPER

THE SIXTEENTH CENTURY

FRANCE, England and Holland are all contestants for the honor of the origination of European wallpaper, with rumors of a challenge from Spain; but the manufacture and use of wallpaper developed so gradually and from so many different sources at once, that it is impossible to assign any one time and place of origin.

Painted papers are the earliest to appear. There is a record of the payment by Louis XI in 1481 to Jean Bourdichon, painter and illuminator, of twenty-four livres (approximately twenty-four francs) for painting fifty great rolls of paper in blue with the inscription *Misericordias Domini in aeternum cantabo* and three angels about three feet high who held these rolls in their hands.

PAINTED PAPERS

In 1507 when Louis XII entered Lyons, painted and gilded paper was used extensively for the welcoming decorations.

In neither of these items, however, does it appear that the painted paper was used as a permanently attached wall covering. A little more significant is an English inventory taken at the Monastery of Saint Syxborough in the Isle of Shepey, County Kent, in the twenty-seventh year of the reign of King Henry VIII —that is, in 1536—which mentions a set of chamber hangings of painted paper. Whether these papers completely covered the wall in the manner of later wallpapers does not, of course, appear in the brief item but the term "chamber hangings" makes this seem probable. The fact that this paper is mentioned at all in the inventory shows that it was considered of some value.

Painted paper appears again in an inventory, this time a French one, that of Guy de Mergey, Canon of St. Peters in Troyes. Two great sheets of painted paper are mentioned showing the Passion and the Destruction of Jerusalem. Again, however, these might have been used as pictures rather than as wallpaper. Jean Nagerel, Archdeacon of Rouen, had a paper of similarly indefinite use that was sold with his effects in 1570. But whatever the use of these particular papers, there was evidently a growing industry in painted wallpapers, for in 1586 in Paris the paper painters' corporation was founded under the name of the "Corporation of Domino Makers, Tapestry Makers and Picture Makers." The term tapestry cannot refer to the real woven material as the tapestry makers were already organized under another name. It evidently indicates the range of patterns the paper painters had come to use, the textile patterns being almost certainly for use as wallpaper.

The domino makers had originally been engaged in the business of copying the marble papers that had been imported into Europe from Persia, which had become very popular for facing book covers and lining boxes and small chests. The term "domino" seems to have been adopted from the Italian, because the first sheets of this paper had been brought into France via Italy where there was a long established commerce with the Near East. From making small marble sheets the domino makers went on to the painting of large sheets of the same designs for wallpapers. The most active center of the industry was Normandy, which was the center also of the manufacture of paper, and of another art, closely related to domino work, *toiles peintes,* painted linen imitating woven tapestry. By 1597 there was a sufficiently large group of wallpaper manufacturers for Henry IV to mention them in an edict; and by 1599 papers for wall covering must have been quite extensively used, for Henry granted a charter in that year to the Guild of Paperhangers.

The oldest fragments of European wallpaper that have been found have all come to light in England. Of these the earliest, from the very beginning of the Sixteenth Century, was found in the Master's Lodge, Christ College, Cambridge. It has a rather large scale pattern adapted from contemporary damask. An interesting Elizabethan paper from the third quarter of that century shows a design in panels of the Arms of England, with Tudor roses, and vases of flowers. Still another paper from about the same period was found in the old Monastery House at Ipswich. These early English papers are all block-printed.

Two block-printed papers were found on the walls of two rooms in Borden Hall, Borden, County Kent, when it was restored a few years ago. These papers seem to date from the latter half of the century, about 1580. Both were fairly well preserved because for over a hundred years they had been protected by wainscot and battening. The more interesting of the two is printed with a small conventional flower design, predominantly in black, on a vermilion ground with the flowers picked out in turquoise blue. The paper is rather heavy and tough in quality. The design is quite definitely Indian, closely akin to the design common on printed cottons of the period.

The earliest specific reference to the manufacture of block-printed paper comes from Holland. In 1568 Herman Schinkel of Delft was tried for printing a book that conflicted with the ecclesiastical law. In the course of the evidence it was mentioned that Schinkel had been printing stripes and roses on the back of ballad paper to be used as covering for attic walls. The fact that ballad paper was used instead of sheets made specifically for the purpose would seem to indicate that this was only a casual experiment rather than the work of an established industry. During this century, however, there was produced in Holland one type of wallpaper that was a regularly manufactured commodity. This was a paper without designs, coated with gold or silver leaf or with metallic paint. There is, however, no record of the exact process.

SEVENTEENTH CENTURY PAINTED AND PRINTED PAPERS

In the Seventeenth Century painted paper continued to be increasingly popular. The commonest designs were diaper patterns, stripes, cartouches and flower designs. Grotesques and cartouches inclosing flowers, fruits, animals and small human figures were also very popular. Sometimes the domino makers became more ambitious, branching out into landscapes and scenes with figures, even, on occasion, attempting to illustrate stories with series of scenes. When, however, they undertook to explain their illustrations with printed titles their ambition overreached itself, for the Letter Printers' Guild brought successful suit against them for infringement of rights and so they had to discontinue this practice.

In none of these patterns did they attempt a continuous repeat. Each sheet was painted separately and formed a separate panel of the design, not connecting at all with the next sheet when they were placed side by side on the wall. Though they were called painted papers most of the French products in the Seventeenth Century were really partly painted and partly printed. The marble papers were, of course, all laid on with the brush, but in the other patterns a black outline was printed with a wooden block. The printing blocks were cut from a design that was first drawn on a large sheet of paper the size of the completed panel, then cut up into squares of a size convenient for the blocks. Very hard wood was used for these, preferably pear-tree wood. In printing, the series of blocks thus made was fitted together and laid face up so that the design for the panel was again continuous as in the orginal drawing. The blocks were then inked with a piece of cloth, a sheet of paper large enough to cover them all was laid face down flat on them and then rolled with a heavy hand roller. This printed the outline. The color or colors were brushed in by hand, water colors or tempera being used. In some cases stencils were cut for each color and the paint put on with the aid of these as in any stencil work.

At first these painted papers, which were probably for the most part rather crudely done, were used only by country people or in the more humble houses in the cities. They were not even sold in shops but were peddled about by hand. But by the end of the century their use had extended even to the better class of houses in Paris, so that Savary des Breslins in his Dictionary that he wrote about 1700 comments on the fact that in the whole city hardly a house could be found that did not have paper on the walls of one small room, at least.

But in spite of the fact that they were so generally used almost no examples of them have come down to us. Because they were inexpensive they were not taken care of and were destroyed and replaced without compunction. Now and then, however, one does find the remains of one of these papers still on the walls of an old house, perhaps the best example being the marble paper on the walls of the older court room in the Town Hall of Veere, Holland.

England seems not to have fostered this

domino work. She turned rather to different types of printed papers. Occasionally fragments of these are found. A number of them are printed in ordinary printer's ink just in outline. Such a paper, for example, is found in the lining of a Bible box from the period of Charles I (1625–49). The design is a coat of arms surrounded by a running vine with leaves and flowers not unlike that on the Borden Hall papers. After the Restoration the use of such papers was greatly extended. Designs continued to be small but the simple outline in printer's ink was often supplemented by one or two colors printed in. On other papers the colors were painted in with oils or with tempera. More elaborate papers were painted with oils, sized with gold and then dusted with powdered colors, giving a rough deep texture.

FLOCK PAPERS

These latter papers were similar in effect to the flock papers that were greatly in vogue at this time. A flock paper, or a *tontisse* as the French called it, is a paper on which the design is printed with some kind of glue and then heavily sprinkled with finely chopped bits of silk or wool so that the material adheres to the pattern, creating a rather successful imitation of damask or velvet. When the glue is dry the extra material is shaken and brushed off and the pattern stands out, well defined by a kind of nap against the plain background, giving quite a rich effect.

This flock technique is quite old, having originated apparently in England. In the Weigel Collection is a Saint George printed in flock, a small fragment that dates from about 1425. The earliest record, however, of the use of this technique for making wallpapers comes from France. In 1620 Le François of Rouen was making flock papers imitating silk weaves for wall hangings. Rouen was, at that time, a great center of the silk weaving industry, so he had any ample supply at hand of the silk waste to be chopped up for his flocks. Le François was succeeded by his son, who continued the business for fifty years, and thereafter for a time by one Tierce, who finally turned from flock papers to flocks on a canvas ground. For a time the firm of Le François enjoyed a large export trade, especially to England, but soon English competition overwhelmed the French production. In 1634 Jeremy Lanyer was granted a charter to manufacture flock paper, and from that time on for more than a century England maintained such supremacy in that field that the paper came to be known even in France as English paper. With an ingenuity worthy of our own advertising-inspired age Lanyer invented a new name for his product, Londrindiana, but the technique was destined long to outlive the name, which perished from disuse.

In Germany wallpapers of all kinds were more reluctantly adopted. By 1670 they were still a novelty, as the comments in a contemporary book prove. When they were used it was only in the women's rooms and even they were more frequently hung with cottons or linens.

CHINESE PAPERS

About the middle of the Seventeenth Century a new influence came into wallpaper decoration. Travelers, merchants and missionaries went to the Orient in increasing numbers and, attracted by the new kinds of art they discovered there, began to bring back quantities of porcelains, glass paintings and other decorations. Among the things they brought were sheets of paper painted with gay designs and made up into sets for the walls of rooms. At first these were regarded only as curios and seldom if ever used as wallpaper, but as more and more of them came to Europe they began to be very popular until by the end of the century they were one of the most fashionable of decorations.

Wallpaper had been used in the homes of China, especially Northern China, since the beginning of the Seventeenth Century. In that century all the decorative arts advanced greatly in China. The new Tartar dynasty of Tsing had come into power and while the rulers of this house did little for the rapidly deteriorating fine arts of the country they did take an encouraging interest in the lesser arts. Especially Kang H'si, second Emperor of the house (1661–1722), fostered the decorative arts, giving tremendous orders to the great porcelain works and taking a keen interest in new glazes and new designs. Wallpaper was one of the decorative arts that rode into favor on the wave of the Imperial interest.

These papers were painted; not, of course, by the great painters but by the artist craftsmen of which there was a very large number in China at the time. Ordinarily these crafts-

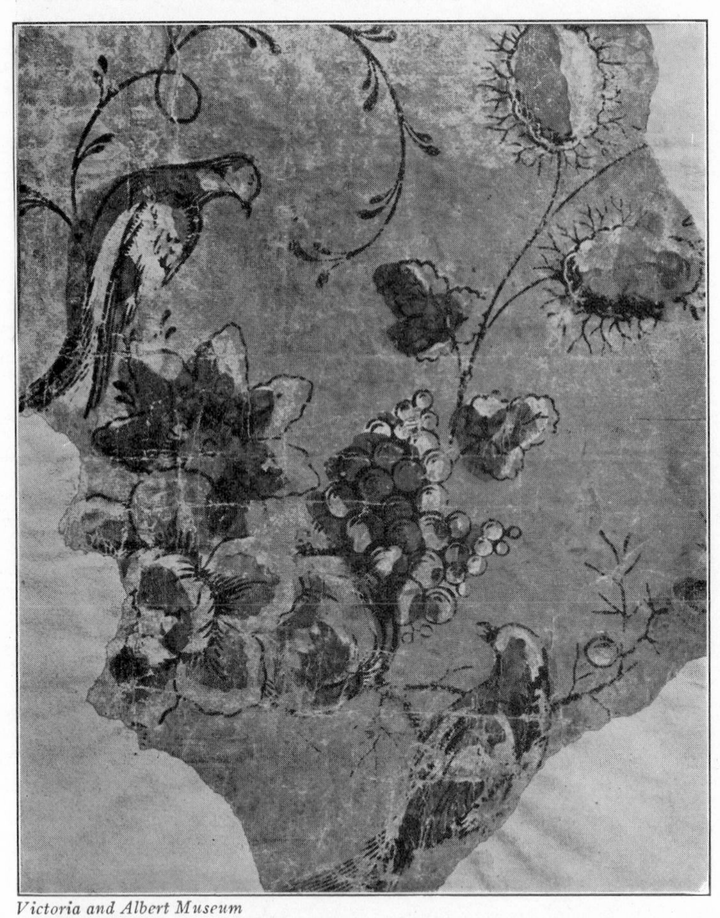

Victoria and Albert Museum

FRAGMENT OF AN EARLY ENGLISH WALLPAPER, SIXTEENTH TO SEVENTEENTH CENTURY.

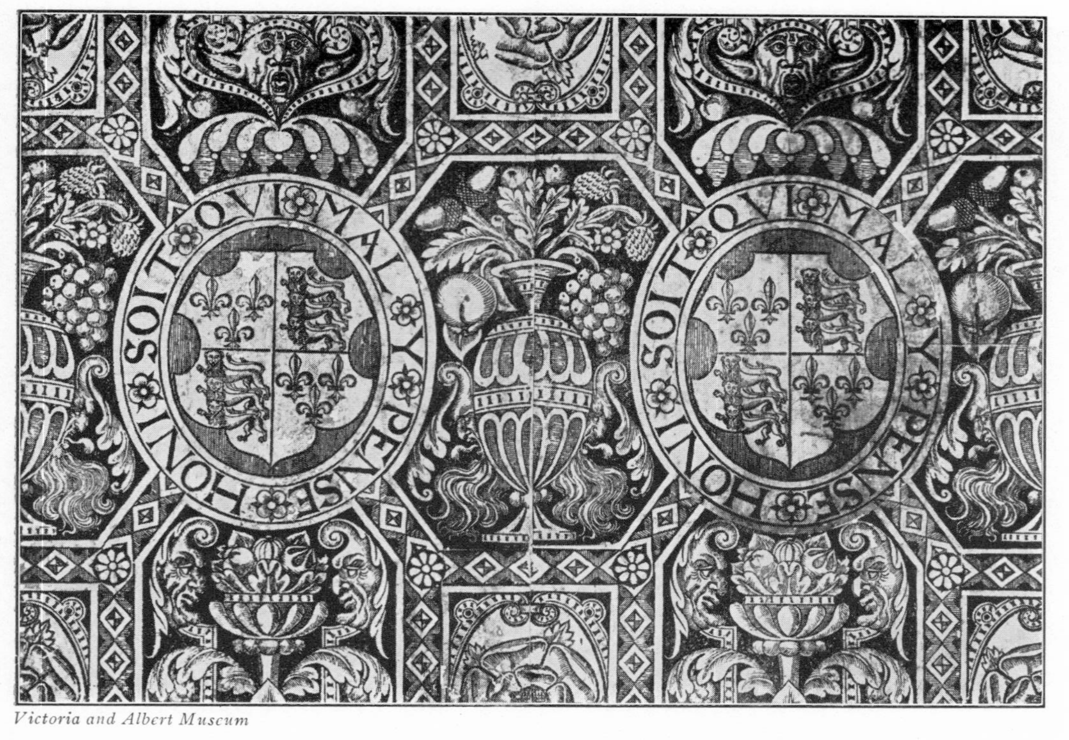

Victoria and Albert Museum

AN ENGLISH WALLPAPER OF THE PERIOD OF QUEEN ELIZABETH.

men had been employed by monasteries and temples to paint religious scenes, altar pieces and banners, or by families to make portraits of their relatives as a commemoration after death. They were many of them well trained men of a good deal of skill but they were not accorded recognition as artists by their countrymen because of the humble, really commercial nature of their work, and the lack of inspiration and idealism in it. Now they added to their other occupations the business of painting sets of papers for the walls of the houses of the wealthy merchants, scholars and government officials.

Their patterns were of three kinds: landscape, bird and flower, and scenes of domestic life. In each of these they were following an established tradition in Chinese painting. In their landscape designs they were the inheritors of a long and honored convention. The word for landscape in Chinese means mountain-and-water, literally, and the painters for hundreds of years respected this literal meaning, combining high peaks with flowing streams and waterfalls in their scenes. In the Sung period (960–1280) landscape painting had reached its greatest height. But these paintings were very abstract, made always with the view to an idealistic and emotional interpretation rather than as the literal representation of any specific place or as a decorative design. During the Ming period landscape painting had become more and more decorative. Into the black-and-white ink paintings of Sung there crept more and more malachite green and strong red, until at the end of the dynasty the decorative conception predominated. The wallpaper painters took these decorative landscapes and adapted them to their own uses, piling conventionalized peak on peak in strong pure greens broken by dashes of black ink indicating trees, and quieter luminous stretches of water. Sometimes, too, they introduced gay little gold and red pavilions and tiny figures of richly robed Mandarins and their ladies. The results were altogether charming, a rich and characterful decoration that took Europe by storm.

The bird and flower designs had almost as old a tradition and a more varied precedent. Certain flowers were especially dear to the masters of Sung. The lotus and the peony were painted over and over again, and the camellia, too, was delicately but strongly portrayed. The birds in which they were interested seem numberless: the kingfisher, the heron, the partridge and many of the brilliantly plumed small birds. During the Ming period, with the increase in interest in the decorative value of painting, bird and flower painting had become more and more popular and more and more colorful so that the wallpaper painters of Kang H'si found many gay and varied models to their hands.

Bird and flower designs had been fostered, too, by Kang H'si's personal interest in porcelains. For it was during his reign that the polychrome wares with their clear strong flower decorations were brought to perfection. Many of the Chinese wallpaper designs were derived directly from the patterns of the porcelains that came in Europe to be known as *famille rose* and *famille verte*. There are in both the same peonies growing out of conventionalized rocks, the same stately roosters, the same cherry blossoms and butterflies and grasshoppers. In some cases the wallpaper painters may actually have copied the porcelain designs, or perhaps porcelain designers copied papers; but probably the similarity is due mainly just to the fact that both derived from the same tradition and were working in the same general established style.

All of these bird and flower patterns are remarkably faithfully and competently drawn. Sir Joseph Banks, the great botanist of the Eighteenth Century, commented on their botanical exactness, and any ornithologist familiar with the bird life of China can instantly recognize the different species shown. Truth was not sacrificed to decoration, but, on the other hand, the decorative effect was never destroyed by the literalness, the most delightful fancy being used in the combinations of the kinds of flowers, their arrangement and the attitudes of the birds and insects.

The scenes of domestic life were of more recent origin. In the Sung period human beings, except in the occasional portraits, appeared only as minor incidents in a landscape. The only exceptions were paintings of famous episodes in the lives of the saints. All painting at that time was strongly religious in character. Even the landscapes and the bird and flower design were intended to convey a universal meaning, to be a presentation of the spiritual essence of the things of this world. Domestic scenes, therefore, were too trivial and personal to be depicted.

But the Chinese of the Ming period, becoming steadily less idealistic in their art, were interested in genre scenes. A whole school of genre painting grew up and innumerable paintings of banquets, festivals, picnics, and even more usual occurrences of daily life have come down to us. The earlier examples of these are all done with great realistic fidelity and many of them are most decorative in their vivid colorings. Later in the Ming Dynasty, however, the drawing becomes less realistic. The figures become abnormally slender, the heads too small, the hands and feet absurdly tiny, the shoulders narrow and round. The human figure, in short, becomes conventionalized.

While this conventionalization of the figures in the domestic scenes was a loss to the art of painting it was a gain to the decorative arts. It delivered these scenes to the hand of the designer ready for an easy adaptation to the needs of his pattern. They were seized with avidity for just this purpose. The porcelain maker put these doll-like people on his plates and jars, the mirror and the glass painters used them for their wares, and the wall-paper painters made them play about the walls of many a merchant and governor.

These papers depicted a wide range of scenes. Sometimes a room would show around its walls the illustration of one continuous narrative. Robert Fortune, the English traveler of the early Nineteenth Century, notes one such paper in the house of a Chinese where he was entertained, the paper of one room telling in detail a long story. In other cases the episodes were not connected, but were simply typical scenes of contemporary life, either indoors or outdoors with a lightly indicated landscape background. But perhaps the most interesting papers were those that showed all the processes of an industry. Of these the most famous in the Western world was that showing the cultivation of tea. The subject seems to have been treated many times and to have been a favorite in the European market.

These Chinese papers fitted in excellently with the decorative needs of the houses of Europe in the late Seventeenth and Eighteenth Centuries. They gave them an individuality and variety and intimacy that was a relief from the rather solemn styles of the Sixteenth and early Seventeenth Centuries. They fitted well, too, into the smaller more informal family homes whose decorative needs were rapidly coming to the fore with the increasing importance of the middle class. Whereas, in the Middle Ages and Renaissance, only the great houses of the nobles and the gentry or public buildings were concerned with decorative effects, because the rest of the world was too poor and too humble to provide for more than bare utilities, by the middle of the Seventeenth Century there had arisen a powerful merchant class congregated in the towns. They did not want vast castles or palaces, but did want beautiful and livable homes and had the money to make them. It was this wealthy merchant class, in direct contact with the Orient through their trade, that used ever increasing quantities of Chinese decorations, porcelains, lacquers, mirror paintings and wallpapers until the Oriental decoration was a raging fashion and wallpaper was firmly established as a central factor in room design.

These Chinese papers were quite expensive even in the Seventeenth Century, and as the vogue for them increased toward the middle of the Eighteenth Century the prices rose. In 1694 a set of six sheets appears in the inventory of Marshall Humière valued at fifteen livres, and while a livre is equal only to a franc approximately, and a franc is normally only about twenty cents, making the six sheets worth only about three dollars, the value of money was then so much greater than to-day the price is much higher than might first appear.

GERMAN GAUFRAGÉ PAPER

Meanwhile there was being produced in Germany an odd type of paper that was destined to leave no trace in the history of wallpaper except a few scraps that have come down to us. This was a paper printed in outline in the textile type of design and slightly impressed into relief, or *gaufragé*. Instead of being printed with wood blocks as all the contemporary papers in both England and France were being printed, this German paper was done with copper plates, a single plate for each design. The plates were warmed before the paper was placed on them and then inked, and the plate and paper together were put through a copper press. The relief impression and the outline were thus registered at the same time, the warmth of the plate

Courtesy E. Weyhe

JEAN PILLEMENT, WHO MADE DESIGNS FOR LE SUEUR, IS ESPECIALLY FAMOUS FOR HIS CHINOISERIES.

Photo by M. E. Hewitt

A FRENCH EIGHTEENTH CENTURY REPRODUCTION OF A CHINESE BIRD AND FLOWER DESIGN.

softening the paper so that the relief impression would be quite strong. Such papers seem to have been made in both Frankfort and Worms between the years 1630 and 1650. Later, in 1670, the first factory in Germany for regular wallpapers was established by Johann Hauntzsch in Nuremberg.

ENGLAND

England, too, in this century tried an experiment in wallpaper printing technique that was to be fruitless of results. In November, 1691, William Bayley received a grant for the sole use of his invention printing all sorts of papers and all sorts of figures and colors with several engines of brass. Why the brass substitutes for the wood blocks were not successful does not appear, but evidently there must have been some defect, for we hear nothing more of them, and English paper continues to be produced in great quantities by the old methods.

FRANCE

It was in France that the wallpaper industry developed most rapidly now under the influence of the increasing popularity of paper as a wall covering. In 1688 Jean Papillon started the first great printing house for wall-papers. His technique was essentially the same as that of the dominotiers who had preceded him, the registering of the black outline with wood block printing and filling in of the colors with the brush either by hand or with stencils. A minor difference was that, whereas the old dominotiers had cut a pattern in a series of small blocks from a paper pattern, Papillon used for his work one plank of wood fully three feet long and sometimes, apparently, drew his design direct on the wood. But these modifications are not of great significance. The important difference between Papillon and the domino workers was that Papillon, for the first time in France, designed and made wallpaper patterns that would join together and be continuous when the sheets were pasted side by side on the wall. In this sense he is the real originator, in France at least, of wallpaper as we know it to-day. His product was so successful that he could afford to maintain a shop instead of peddling it from place to place as the old painted papers had been sold.

LUSTER PAPER

Papillon has to his credit, too, another invention in wallpaper, the so-called luster paper, though similar paper had long been made in England. In this paper the design, instead of being painted or stenciled in colors, was powdered over with ground-up paints. Papillon's grandson gives the following recipe for its manufacture:

"To 2 ounces of fish glue heated lukewarm and melted, add twice the amount of starch well diluted; stir until there are no lumps and it is all well mixed. Let the mixture stand until the next day and when it is to be used heat lukewarm. Then, having sketched the design in lightly with charcoal, paint it with this paste; then scatter on it ground-up paint of the desired color and let it dry so that when the sheet is shaken the powder will remain only on the design."

The process could also be applied to a design of several colors by painting each part of the design for each color separately, scattering that color on it, leaving it to dry and then repeating with the next color. Powdered metals could be used also in place of the colors. This paper was considered a great improvement on the domino papers because of the greater depth of the colors and the stronger quality of the texture.

EIGHTEENTH CENTURY FRANCE

Papillon's device of the continuous repeating design was immediately imitated by a host of followers. Most of these were engravers by profession, map makers and illustrators of small importance who saw in the new field a lucrative trade. The first of these was Adam, who set up his shop in 1700. With him was Blandin, who made architectural designs for him. The cleverest of the imitators was Le Sueur, who had been an apprentice with Papillon and then set up as a rival with Vincent Pesant, Blondel and Panseron as his designers. Pierre Panseron, who was a pupil of Blondel, had the advantage of being not only an engraver but a practical architect. After acting for a time as a teacher of drawing in a military school, he became superintendent of construction for Prince Conti and wrote many books on architecture which he illustrated with his own engravings.

Other engravers who were in the business of either printing or designing wallpaper about this time were Roumier, who, from 1727 on, designed a number of fine great plates of flowers and ornaments; Dufoucroy, who was on rue Jacob-Saint-Germain; Jean Pillemont, who had made designs for the Gobelins and very famous patterns of Chinoiseries with flower gardens and trellises for the silk makers of Lyons, and who later became court painter for both Poland and Marie Antoinette; Masson and his successor Miyer; Basset, Forcoy, Vaseau and Goupy. Breton, both father and son, continued to make marble papers. Dominotiers who were still producing in the first half of the Eighteenth Century included Letourny on rue d'Orléans, Basset and Rabier in the city of Orléans.

The business of Jean Papillon had meanwhile several difficulties to meet. In 1708 he was threatened by the Painters' Guild for infringement of their rights; but he succeeded in avoiding legal difficulties. In 1723 Papillon père died and left the business to his son. His son had from earliest childhood been forced to put aside his ambition to become an engraver of more dignified subjects and obliged instead to cut wood blocks, design papers and even to go out and hang them in the customers' houses. Consequently he did not feel very enthusiastic about the business. In 1740, therefore, he sold out to the Widow Langlois, who in turn left it to her son in 1766. While the business was in the hands of Papillon's son it had met and overcome further difficulties in the shape of threats from the copper engravers that some of their rights were being infringed. Still another difficulty now arose in the plagiarism of one of Papillon's old workmen, Didier Aubert, who went into producing on his own account but used the name Papillon.

ENGLAND

In spite of the success of Papillon's papers, they still could not displace the English papers in public favor. The English papers now included both the block-printed papers and the flock papers that had been manufactured there since 1634. The latter were usually in damask designs and in one color, generally blue, or one color on a gold ground. The block-printed papers of England differed from the French papers in that not only the outline but all the colors, too, also were printed in with wood blocks. This was done by cutting a series of blocks for every part of the design, which had first been drawn on paper, and painted in water colors or tempera, the first block being for the outline, the second having in relief the parts to be done in the first color, the third for the next color and so on until every part of the design was represented by at least one block in relief for every color. The parts of the design that were too delicate to be cut conveniently and safely in the wood were laid in with lines and pegs of brass. The background was put in with a brush. The outline was then printed. When this was dry the first color was printed with the appropriate block, the other colors being printed in order either from light to dark or, if a bright clear effect was desired, an effect more nearly resembling fresco, the order was reversed and the printing done from dark to light, leaving the high lights clearer and more sharply defined.

In this process the block is laid on top of the paper instead of the paper on the block as Papillion had done. The block is then pressed down by hammering with a mallet, rolling with a heavy roller, or rolling between two rollers on a frame, a simple type of machine used by the old type printers. The colors used were very heavy and pasty, usually thickened by an admixture of glue, and were entirely lusterless. The designs were sometimes in cameo but more often in full colors and included flowers, damasks and all types of ornament.

Credit for inventing this method of printing was claimed by John Baptist Jackson, who had a factory in Battersea. Jackson, who was born in 1701, had been apprenticed to the engraver Kirkall. When he was a young man, about 1726, he went to Paris and worked with Papillon fils, who, although he had sold his father's wallpaper business still taught wallpaper engraving. It is probable that Jackson learned the rudiments of the art here but he did not begin to practise it at once. Instead he went into the more ambitious field of engraving reproductions of the Old Masters, including one of Rembrandt's Descent from the Cross and a series of seventeen after Titian and others that was published by Pasquale in 1745. Perhaps this kind of work did not pay or perhaps he was restless for further experiment. Whatever his motive may have been, the next year, 1746, he entered a factory in Battersea to make wallpapers and eight years later, 1754, issued a book on the subject. It is in this book that he lays claim to the invention of chiaroscuro

Metropolitan Museum

THE BIRD AND FLOWER WALLPAPER DESIGNS OF CHINA WERE DERIVED FROM AN OLD TRADITION IN CHINESE PAINTING.

Photo by M. E. Hewitt

THE LANDSCAPE DESIGNS OF THE CHINESE WERE BEGINNING TO BE VERY MUCH IN DEMAND IN THE REIGN OF QUEEN ANNE.

printing in wallpapers, as he himself calls it. The book, however, is not conspicuous for the modesty of its tone, being frankly a self advertisement fishing for a patron to promote his business as the Duke of Cumberland had promoted the tapestry works at Fulham. The claim, therefore, may be part of the exaggeration usually felt to be necessary in advertisements. Or his invention may consist merely in substituting oil colors, which he especially emphasizes, for the customary water colors mixed with sizing. It is, however, clear that not all English papers of this time were being printed in color, for Horace Walpole in a letter dated 1754 speaks of having his paper painted after it was hung, but from contemporary advertisements it appears that painted paper did continue in use after the invention of color printing, so this is no reason to accept Jackson's claim.

Jackson of Battersea did not use his method to make the continuous repeating patterns that Papillon had popularized in France. Jackson's designs are all for panels to be used in elaborate schemes, with numerous and varied borders adapted to the spaces of the particular room. He is especially proud of his relief effects, suggesting that his prints of antique statues in niches are a worthy substitute for real sculpture for those who cannot afford it. He also suggests paper imitating stucco which he claims to be able to reproduce most effectively. In fact, his offering is all-inclusive: "The Appolo of the Belvidere Palace, the Medicean Venus and other antique statues, landscapes after Salvatore Rosa, Claude Lorraine, views of Venice by Canaletti, copies of all the best painters of the Italian, French and Flemish schools, in short, every Bird that flies, every Figure that moves upon the Surface of the Earth from the Insect to the Human, and every vegetable that springs from the Ground, whatever is of Art or Nature, may be used for fitting up and furnishing rooms." The book is illustrated with small color engravings of statues, architecture, birds and so on as samples of the possibilities.

It is appalling to imagine the effect of these papers in place and Jackson's contemporaries did not all welcome them with unqualified approval. Horace Walpole speaks of not being able to endure the Venetian prints while they pretended to be copies of Titian. In the end, however, he did succumb to them and used them on his walls. In a letter of June, 1753, he says, "When I gave them the air of barbarous bas-reliefs they succeeded to a miracle. It is impossible at first sight not to conclude that they contain the history of Attila or Totila done about the very era."

It was probably Jackson of Battersea's papers that were referred to in the following advertisement from the London *Evening Post* of January 8, 1754: "The new invented paper hangings for the ornamenting of rooms, Screens andc., are to be had by the Patentee's direction of Thomas Vincent, Stationer, next door to the Wax-work in Fleet Street. *Note.* These new invented paper hangings in Beauty, Neatness and Cheapness infinitely surpass anything of the like nature hitherto made use of, being not distinguishable from rich India paper and the same being beautifully colored in pencil work and gilt." There follows a warning against infringement for fourteen years from August 1753. "India" was the popular trade designation for the Chinese imported painted papers.

By 1750 the English technique of multi-color printing with wood blocks was brought to its utmost perfection by George and Frederic

Eckhardt, who had a shop in Chelsea. Here they printed not only papers but silks and linens, too, using the same patterns interchangeably for all three materials.

Of the other English paper makers of the time we know little. Masefield, who had a factory on the Strand about 1755, claimed that he too had a method peculiar to himself "which surpasses anything of the kind yet attempted." He advertised Landscapes, Festoon and Trophies, India paper and Mock India paper. Matt Darley, who engraved some of the plates in Chippendale's Director, was located at the Acorn facing Hungerford, the Strand. He had "Ceilings, Panels, Staircases, Chimney Boards and so forth, neatly fitted up either with paintings or stainings in the Modern, Chinese or Gothic Tastes for Town or Country." He offered "large allowances for ready money." The "Modern Taste" perhaps refers to the floral papers, one of which Horace Walpole describes in a letter of 1753—"a blue and white paper in stripes adorned with festoons."

English papers had always been more highly prized in France than the local productions but by the middle of the century, with the incoming wave of Anglophilmania that influenced all the fashions in dress and decoration, it became a veritable rage. In 1754 Madame de Pompadour herself ordered an English paper for her dressing-room at Versailles, and in 1758 put English paper in the bath at Château de Champs. Flock and especially block-printed papers for the time superseded tapestries for wall decoration.

ENGLISH METHOD IN FRANCE

France herself had by degrees been acquiring this English technique. By 1735 Simone was making landscapes and verdures in the flock work, M. N. B. Poilly had begun, also, to make flocks and in 1730 Boulard had started the production of block-printed papers in which one color was blocked in, a method immediately imitated by shops in Chartres and Orléans. But none of these enterprises had made any great impression on the market. In 1750, however, France began to forge ahead in this branch of the art. Jaques Chauvan, a former apprentice under Papillon, invented an improved method in which he printed in the colors with blocks using, however, not the sized colors commonly employed by the English workman, but oil colors. The oil colors had the advantage of resisting moisture but did not give the heavy lusterless surface so much prized. In the same year Jaques Gabriel Huquier set up a plant for imitating exactly the English papers. Thereafter this method completely superseded that of the domino workers and of Papillon.

The industry now began to be sufficiently important to be regulated by ordinance and protected by taxes. In 1760 a law was passed establishing nine aunes, that is about 34 feet, as the length of a roll of wall paper. The use of rolls in place of the short lengths formerly employed had only recently begun, Fournier having been the first experimenter to try putting sheets together to make a continuous length. In 1765 the French manufacturers were protected from outside competition by a tax of 100 sous the hundred-weight on imported papers, while the export tax was only 10 sous per hundred-weight.

France had no sooner taken over the English method than she began to improve on it. The English flock papers had always been printed all in one color on either a ground of the same color or on a gold ground. In Lyons, Lecomte, in the early 1760's began to print flock patterns in polychrome, dusting on chopped silk of one color on one part of the pattern and letting it dry, then doing the next part in another color and so until he had a very effective imitation of the elaborate and vari-colored brocades of the period. These brocade flock papers were very much admired, apparently with good reason, judging from a contemporary description. Unfortunately none have come down to us, probably because they were rather perishable. The *Literary Year* of 1764 says:

The richness and elegance of these papers is admirable. The colors are very varied and are shaded with great art. In one are bunches of roses and all kinds of flowers that give the most delightful and splendid effect. In another on a striped background are scattered little bouquets between which a ribbon is thrown which ties together all the objects and is unfolded with real grace. In a third there are Chinese cartouches with figures giving a concert. There are others more appropriate for dressing-rooms or boudoirs in the Chinese taste, and others in grisaille with baskets and garlands of flowers arranged in the most graceful manner. The price of the papers is fixed and one cannot bargain for them. The Chinese and that with the ribbons is 25 cents for 46 inches (25 sols per aune); the rose vine on a satin ground is 35 cents, on a moiré ground 40, the Chinese concert 45 cents. The size of the rose vine design is 15½ inches (15 pouces, 5 lignes) that of the design with ribbons 19¼, and the others 20. These papers take the place of materials at from 60 to 80 francs for 46 inches and if the materials do last, the colors fade at least as quickly as the colors of the papers, and when a material is faded it matters little whether it is worn or not.

The patterns described here are characteristic of this Louis XV period. Stripes were omnipresent, always undulating in full curves, simulating in texture and design ribbons, passementerie, lace and even fur. Moiré silk was newly contrived and very popular. The stripes were scattered with small bouquets of flowers and little bows and knots of ribbon. Later in the reign, the stripes became more purely floral indicated by a continuous chain of interwoven branches with their leaves and blossoms. The colors were those introduced by Boucher in his painting, high in key, but delicate and very vivacious—dove gray, rose, pale blue and light leaf green.

It was at this period that Chinoiseries, too, were at their height. These quaint and frivolous designs of bal masque Chinamen as they were conceived by the irresponsible imaginations of the French court were specially favored by Madame de Pompadour. Under her influence Boucher painted fantastic car-

Metropolitan Museum

THE CHINESE LANDSCAPE PAPERS WERE ADAPTED FROM THE DECORATIVE LANDSCAPE PAINTINGS OF THE MING PERIOD.

Photo from M. E. Hewitt

MANY OF THE PAPERS SHOWING SCENES OF DAILY LIFE ARE VERY SIMILAR TO THE MIRROR PAINTINGS THAT WERE IN VOGUE AT THE SAME TIME.

toons of them for both the Gobelins and Beauvais tapestry factories, and his lesser followers hastened to imitate. Madame de Pompadour's championing of the Chinoiseries may have been due in part to the fact that she was a patron of the French East India Company and saw here a chance to further their interests, for the Pompadour was a shrewd and active politician, but fundamentally the reason for their success went deeper than this. The French court in its passionate pursuit of pleasure was constantly threatened by ennui, the *bête noire* of hedonists. They found an antidote in the unfamiliar and uncertain romance of strange lands, and in a make-believe life patterned on that of these strange lands, which they substituted for the real life that was palling.

The French interpretation of the Chinese in wallpaper patterns borrowed little or nothing from the real Chinese wallpaper that showed landscapes and birds and flowers. Perhaps these papers still conveyed enough of the idealism of their old tradition, even in spite of their degeneration into decoration, to keep them immune from frivolous parody. But the French designers did take hints from the domestic scenes, tricking out the doll figures the better to suit their fancy. It was the lacquer patterns, however, that offered the most suitable material, with bits of silhouette landscape and architecture framing gay little silly mannikins. With these as a starting-point the French pattern makers created a world of diminutive Chinese such as never existed before in any land, garbing them in foolish, amusing distortions of true Chinese costumes. They set them down in strange scenes showing the wonders of India, China and the equally alluring tropics; scenes compounded of vague reports, fancy and imagination, for almost none of the designers could have had first-hand information of these regions. With delightful incongruity, they did not hesitate even to mingle their Chinese with pure French motives, standing, perhaps, a fancy-dressed Chinaman on a frilly lace stripe. The variations on the theme seemed almost inexhaustible and they were produced in great quantities up to about 1780, when the vogue began to wane. Some of the patterns, however, continued to be printed well into the Nineteenth Century and there has recently been a strong revival of them.

Lecomte's polychrome flocks marked a high level both in design and in technical achievement. The shading of the flock colors and the variety of textures produced, silk, satin and moiré, both indicate a great advance over the English productions. Unfortunately, however, the factory did not last long; in 1769 Lecomte died and left it to his widow who soon gave it up.

It is not surprising that bargaining about the prices of Lecomte's paper was out of the question, for they were relatively low. In 1749 Lady Mary Wortley Montague, writing to her daughter in England from Louverne, said: "I have heard of the fame of paper hangings and had some thought of sending for a suite, but was informed that they were as dear as damask, which put an end to my curiosity." Perhaps Lady Mary was accustomed to English prices, which seem to have been lower, for Lady Hertford speaks of having seen papers at twelve and thirteen shillings a yard, and others at four shillings, and having finally bought one at eleven pence. This would make the paper about twenty-nine sous an aune, roughly calculated, as compared with the Lecomte papers at twenty-five to forty-five.

There probably was, of course, a difference in quality. The Chinese papers, which were now being imported into France direct by the French East India Company, had risen greatly in price. In 1655 six sheets showing landscapes and figures sold for 144 livres—that is about 144 francs or approximately $28.80, if the difference in the purchasing power of money is not taken into account. Another set of Chinese paper, twenty-four sheets with gilt ornament and figures, each ten by three and a half feet, was advertised for sale by M. Martin in a Paris paper of May 8, 1770, at 24 livres —about $4.80—a sheet, to be sold all together, or in sets of eight sheets each.

All the manufacturers at this time sold their own wares at retail, usually at the same address at which they maintained their workrooms. But there were also a few retail dealers who did not manufacture. The industry had gone a long way from the days when the painters themselves peddled their sheets from door to door. Among the lesser manufacturers who had shops at this time in Paris were Roguié, who in 1754 was on rue du Cloître-Saint-Germain; Aubert on rue St. Jacques, the center of the engravers' studios, at the Sign of the Butterfly; Garnier, who in 1762 was on rue Quincampoix, at l'Hôtel de Mantoue; Jacques Chereau, who already in 1740 was making experiments in polychrome printing; Pancet and Daumont, both of whom were producing in 1770. Among the retail dealers were Niodot, who had the Song of the Lark on the old Place du Louvre in 1769; and Crèpy the Elder and Watin, who in 1770 were both in rue Saint Apolline. Other manufacturers and dealers were Lancake, an Englishman who in 1769 established a factory for wallpapers made in the English manner at Carrière with a store for selling his wares on rue Geoffrey Lanier; the Demoiselle Hennery, who in 1774 was on the rue Comtesse d'Artois; Mathon, whose shop was running in 1777; Windsor, who in 1779 advertised papers imitating sculpture and architecture from a shop on rue Petit Vauguard; and Damier, who had been on rue Dauphine in the Hôtel de Genles but moved in 1780 to the Hôtel de la Grenade on rue de Bussy. Outside of Paris, Marseilles, Besançon, and Bordeaux all had factories using the English method.

Meanwhile the method of hanging wallpapers had undergone several changes. The earliest paper found in place, that of Borden Hall, 1580, was apparently tacked to the plaster that filled the spaces between the beams and the uprights. The usual method in the early houses, however, seems to have been to set up a wooden frame stretched with canvas over the brick or stone walls so that there was a slight air space between the stone and the canvas. The air space helped to keep the paper dry and so to protect it from rapid disintegration. By the Eighteenth Century it was customary to paste the paper directly on strips of canvas and then tack or paste the canvas to the plaster. Occasionally in place of canvas thick rice paper from China was used for this purpose. Strengthened with this lining, the paper could even be removed from the house if necessary. There were for sale quantities of paper already lined that had been used, and occasionally today sets that have been preserved in this way come to light. But by the middle of the century the practice of the canvas backing was being discontinued. Mrs. Delaney in a letter dated June 1750 comments on this. "When you put up paper," she says, "the best way is to have it pasted on the bare wall. When lined with canvas it always shrinks from the edges." The practice of varnishing paper did not become general until later—about 1780.

Holland about this time, 1768, was producing a high-class wallpaper covered with gold and silver. The manufacturer of this was Eccard, in The Hague. Twenty years later gold and silver papers were made in Germany, in Bavaria, after a process that was introduced there in 1781 by a Frenchman named de Couvier.

About this time was established one of the famous factories of France. Reveillon, who in 1759 had been a paper merchant on rue de l'Arbre Sec at the Armes de Conty, sometime between 1770 and 1775 started a factory for wallpapers in the Faubourg Saint Antoine with a shop that was both wholesale and retail on rue du Carrousel, opposite the entrance of the Tuileries. The factory grew rapidly and was soon the largest in France, with three hundred employees. It was so successful that in 1784 the title of Royal Manufacture was conferred on it.

Reveillon had the good judgment to surround himself with excellent designers who produced a very wide range of patterns.

The discovery of Herculaneum in the middle of the century and the publication of several books of designs from the wall paintings there had brought into fashion Pompeian designs. Reveillon specialized in a type of paper that imitated these frescoes very successfully, J. François Van Dall of Anvers designing most of this type of papers for him. Papers introducing architectural ornament were also greatly in vogue, and with the growing classicism there was a larger and larger demand for papers reproducing Roman ornament. Percier designed some famous antiques for him and the Van Spraendonck Brothers specialized in designs of marble vases on consols with bas-reliefs and medallions. Salembier contributed skilful arabesques. Flower designs continued to be in favor, so Reveillon kept able flower painters like Joseph Lauront, Mèry père and Huet on his staff. Others designers who worked for him from time to time were J. J. Fan; Cietti or Sietti the Italian; Prieur, one of whose most famous papers was the Vendanges; Paget; Lavallé-Poussin; Cauvet; the Rousseau Brothers; Charles Monnet; Joseph Sauvage of Tournai; and Boissellier. In addition to the Pompeian and classical designs the

ONE PANEL FROM THE LIFE OF PSYCHE DESIGNED BY LAFITTE AND PRINTED BY DUFOUR IN THE FIRST QUARTER OF THE NINETEENTH CENTURY.

Courtesy Charles Grimmer and Son

THE IRRESPONSIBLE FRENCH INTERPRETATIONS OF CHINESE DECORATION WERE REPEATED WITH MANY VARIATIONS FOR ALMOST A CENTURY. EARLY NINETEENTH CENTURY EXAMPLES.

usual patterns of the period were those characteristic of all the fabrics of the reign of Louis XVI. Patterns were very small and often scattered with flowers in bouquets, in baskets, in vases, and various ornaments such as ribbon bows, medallions and gardening tools; these last in commemoration of Marie Antoinette's elaborate pastoral affectation.

Although by degrees the technical difficulties were being overcome, the wood-block printing was still at this time not quite perfect. After all the colors had been printed, it was still necessary to go over the whole design and touch it up by hand with a brush. Nevertheless very elaborate designs were produced, Boissellier having designed one for Reveillon with twenty-four colors. The greatest problem, however, was in finding the proper paints. The Chinese papers had absolutely insoluble colors but in the European papers the colors had always had a tendency to run when the wet paste was applied. Finally, in 1783 the Duc de Chaulnes perfected an insoluble dye that enabled the French manufacturers to use a great number of colors without danger of the effect being ruined when the paper was moistened. Reveillon had some difficulty, too, in getting the right quality of paper he felt he needed. To solve this problem he bought a paper factory at Courtalin-en-Brie and manufactured his own. His product was of such high quality he was awarded in 1785 the prize for the Encouragement of the Useful Arts founded by the famous financier Necker.

Reveillon's factory produced a great number of papers. They were rather expensive, and necessarily so, because of the number of workmen required to maintain the plant and the growing elaboration of the processes, but nevertheless they found a wide market. The large overhead, however, worried Reveillon and he made an attempt to reduce expenses. But the attempt was made at an ill-chosen moment and in an unwise way. In April, 1789, he announced a cut in half of the wages of all of his workmen and refused to pay any attention to their protests. The Revolutionary spirit was rampant and in a moment his group of three hundred employees was turned into a mob shouting for vengeance and destruction. That day the mob was dispersed but the next day it returned with undiminshed fury, destroyed the factory and ruined and burned most of its property. Reveillon abandoned the enterprise and fled to England.

Reveillon himself in a defense written during his exile denied that he had tried to underpay his workmen, emphasizing the fact that he had himself started life as a workman and so understood and sympathized with their difficulties, and citing his payroll in evidence. His engravers and designers got from fifty to one hundred sous a day, his printers thirty to fifty, his porters and general laborers from twenty-five to thirty and the children from twelve years old up from eight to fifteen. Painters who worked by the piece earned as much as six to nine livres, roughly equivalent to francs, a day, and certain painters got special bonuses of from twelve hundred to ten thousand livres. He attributed his troubles to jealousy aroused by his prestige.

Two enterprising Englishmen took advantage of the destruction of this famous house to create a business for themselves. Their opportunity was the greater because the war had cut off all importations from England. These Englishmen—Robert, an English merchant in Paris, and Arthur, an English clockmaker, who had both had some training in the wallpaper business under Reveillon—set up a shop on the Boulevard at the corner of rue Louis-le-Grand. They specialized in sepia and grisaille prints and built up a successful enterprise. Among their most important productions were panels with architectural frames and engravings after famous paintings. Among others they reproduced in this way several different paintings by Boucher of amorini, works by Delafosse, Van Loo and Fragonard, and landscapes by Hubert Robert. Their best engraver was Ridé. During the Reign of Terror Arthur was guillotined and the firm was then carried on by his partner under the name of Citizen Robert on Place Vendôme.

Two years after the destruction of Reveillon's the firm was revived by two successors, Jacquemart and Bénard. Like their predecessor these manufacturers employed the finest designers. Prieur, Huet and Percier continued to work for them, and Boucher fils, Guérin, Costain, Brocq and probably Fontaine were all on their staff. But the kind of design in vogue was changing. Jacquemart put out a number of papers appealing to the popular taste by the use of Republican emblems, and he introduced, also, the papers imitating draped fabrics that were much in vogue for the next forty or fifty years.

Any paper imitating a fabric was a success at this time. Chouard of Lyons in 1788 introduced one that looked like muslin and for some years there had been in use designs simulating gathered satin.

Another flourishing house of this time was that of Legrand. He had a factory on rue d'Orléans and a retail shop on Place Dauphine. It was he who issued Andouin's Pompadour paper in 1797, some of Mèry père's floral papers and Legendre's medallions and frescoes.

Another post-Revolutionary enterprise was that of Bellanger, who undertook to found two factories. One was with Dugourc. They specialized in papers with Republican designs and in patterns in the antique manner, and no less a person than Prud'hon, then impoverished by the political upheaval, made designs for them. For some reason, however, they soon went out of business. The other establishment was with Amisson Duperron and this one continued to produce.

In 1797 there was founded the house of

Zuber, publisher of many famous papers and still producing. He established himself at Rixheim near Mulhouse, where the factory still exists. Among his more notable designers were Mongrie, Rugendas, Hermann, Chabal, Dussurguey and Dumont. Perhaps the most prominent and successful of them was Laurent Malaine, who made flower patterns. He was a skilled flower painter and had been a designer for the Gobelins and later for some of the silk makers of Lyons.

THE NINETEENTH CENTURY. FRANCE

In 1803 Zuber began to publish a series of great panorama papers, the first of which, scenes in Switzerland, was painted by Mongin. Most of the papers of this series were issued about 1830 or immediately thereafter and were designed by Ehrmann, Zipelius and Fuch. There were scenes from all parts of the world, tropical forests, mountains, seaports, with and without figures and incidents. One of the most noted and enduringly popular is Isola Bella, a rich fringe of tropical plants silhouetted in many clear colors in the immediate foreground against a distant vista.

An establishment that was to become one of

Zuber's strongest rivals was opened a few years after Zuber's plant. This was the house of Joseph Dufour. He had already been in the business at Macon but early in the Nineteenth Century moved into Paris and established himself at 8 rue Beauveau, where he remained until 1845. He secured a number of able advisers and designers—Laffitte, Madèr père, Madèr fils, Wagner, Portelet, Délicourt, Fragonard fils, and Alexandre Evariste, who was a pupil of David and so began his career as a Classicist but later turned Romanticist. Dufour was especially noted for his grisaille papers, the most famous being the Cupid and Psyche series after a design by Laffitte, the cartoons of which were painted by Madèr père. It shows the story of Cupid and Psyche in the version then most popular, that of Fontaine. It was first issued between 1808 and 1814. Macon also did an amusing design for him in 1807 that became quite well known, a landscape with Indians and Spaniards figuring in various episodes. This was, of course, a product of his romantic period. Later Macon came under the influence of Percier, who was entirely controlled by David and so turned Classicist. Other famous publications of this house are the Adventures of Telemachus, especially noted in America because of several fine early examples to be seen here in old houses, and the series of the Months by Fragonard fils. Dufour was succeeded by his son-in-law, Leroy, who continued to publish many of the old designs including the Telemachus and the Greek fêtes by Madèr and who issued also the Incas of Maçon, and Paul and Virginia which Brocq painted.

Still another firm of the early Nineteenth Century was that of Dauptain on rue Blanche, Mibray. When he died in 1811 the work was carried on under the direction of his widow and son. His most noted designers were Martin Polich, Aimé Chenevard, and Portelet. The house issued a wide range of designs including Moyen Age, Renaissance, Rocaille and Pompadour patterns. One particularly interesting paper that he published illustrated Molière's "Precieuses Ridicules."

The house of Semon in the Jardin des Capucines was founded in 1810. In a few years his son succeeded him and took Cortulot as his partner. This venture lasted only twenty years, going out of business in 1840.

In 1821 was founded another one of the

firms destined to survive through a long and productive career. The originator of this was Madèr fils, who already had had experience in the business as a designer for Dufour. Another Dufour employee, Délicourt, who was familiar with the technique of production, joined him. When Madèr died, Délicourt took over the entire management for Madèr's widow. In 1834, however, Madèr's sons were ready to take up the work so Délicourt retired from the firm. In 1849 the business was sold out to Défossé. Later he took as partner Karth and the business is still conducted under the name Défossé and Karth.

This firm has issued many famous designs including the Prodigals by Thomas Couture.

When Délicourt left Madèr's business he set up for himself, but the enterprise did not last long. In a short time he took two partners, Campas and Gurat, and in the end sold out to Buzin. For designers they employed Martin, Wagner, Riesner, Dumont and Ch.-L. Müller. In 1860 they went out of business. But in spite of the short duration of this firm it managed to make a very interesting contribution to wallpaper design. This was a series of panel papers some of them so contrived that the dimensions of the panels could be adjusted to the proportions of the room. In l'Elysée, for instance, each landscape is framed by an architectural structure overlaid with a trellis and a blossoming vine that can be moved in or out to fit the paper, both in height and breadth, to any particular wall. Another panel paper of Délicourt's, which is not, however, variable in dimensions, is that of the Sciences, a very rich design showing allegorical, pseudo classical female figures in a setting of arabesques, flowers and amorini, the two large panels having each Astronomy and Geometry with Physics and Chemistry in subordinate positions. The most famous Délicourt paper was the great Hunt after Desportes.

Other manufacturers and dealers in Paris in the early Nineteenth Century were Damiens in rue de Bussy, Montrille in rue Vivienne, Jacques Albert, 15 rue du Bac, Gaguet and Caffère, 4 Place Vendôme, who specialized in satin papers with a silver luster, Dodard at 25 rue Ferdinand, Legendre on rue Pâte-Saint-Antoine, Cartulat on rue Napoléon, Paulot and Carré, 5 rue de Reuilly, the Frèsnard Frères, Périgueux, Vitry, Masson and Chicaneau,

Vélay, 10 rue Leloir, Vauchelet, Boulanger of rue Saint Benôit and J. Guillot. On rue Charleton an Englishman, Henry William, set up a shop in 1825 where he specialized in those imitations of wood and marble that were the beginning of the end of the artistic value of wallpaper. In Besançon, Bourier was producing; in Nancy, Langier and Conolis; in Caen, Le Flaguais, father and son; in Lyons, Pignet; and in Saint Denis, Richon, who had two factories. In Vienna there was the house of Mognat-Perrin and that of Wery.

A quaint product of the early Nineteenth Century was the commemorative papers. One in honor of Washington's death was issued, appropriately solemn. It showed a repeating design of a tomb inscribed "Sacred to Washington" surrounded by an iron fence and framed by columns and an arch surmounted with an urn and a mourning eagle. Liberty and Justice, likewise Mourning, stood at either side. In front were crossed arms and flags. It was printed in gray and black to complete the appropriately depressing effect. Another particularly interesting paper obviously intended for the French royalists, commemorated the Battle of Waterloo. The ground of this is pale blue with white stripes indicated by a design of a running vine intertwined with the continuously written names of all the generals of Waterloo written in script in capital letters. The initials of the reigning monarchs of the countries involved are in wreaths, while printed in gold across the top is a border of military medals.

ENGLAND

In England another amusing wallpaper fad had arisen. The paneled wall had always been in favor there, Jackson of Battersea's papers, for instance, being all in panels, and at the end of the Eighteenth Century the fashion had recurred with renewed force. Sheraton lent it additional prestige by including several room designs in his pattern book that showed paneled wallpapers. Now paneling wallpapers necessitates, of course, the use either of architectural moldings or of borders. The prevailing preference at that time was for the paper borders. It was in the designing of these borders that new and individual styles were attempted. Thomas Rowlandson, the noted English caricaturist, engraved one set of these borders after drawings by G. M. Woodward: The design, if design it can be called, is a succession of entirely unrelated sketches with about an inch of blank space between every two drawings, each with a caption composed of a quotation or a bit of dialogue. There is no connection between the incidents illustrated nor even a continuous pattern along the edge to hold the sequence together. The caricatures are in Rowlandson's typical manner, colored with the usual pale blue and pink washes. The borders were printed on sheets, three borders to a sheet each about three inches wide and a foot and a half long. In using them they had to be cut out and joined end to end. In the complete set there were twenty-four sheets enclosed, loose, in a folio. The book was published by Rowlandson's usual publisher, R. Ackerman, 101 Strand, in 1800, under the title, "Grotesque Borders for Screens, Billiard Rooms, Dressing Rooms, etc., etc. Forming a caricature Assemblage of Oddities, Whimsicalities and Extravaganzas. With appropriate labels to the Principle Figures." Amusing as these borders are, as a room decoration they could not have been an unqualified success, for at a very short distance drawings and legends alike would be undecipherable, so that the effect would be only of a series of pale detached blots.

More successful from a decorative standpoint was another similar folio engraved by Merke after R. B. Davis and published by C. Random at the Sporting Gallery, 65 Pall Mall, London, in 1810. These borders, similarly printed in sections eighteen inches long and five and three-quarters wide, were entitled "New Invented Borders for Rooms etc., of Field Sports," and showed various scenes of kennels, stables, hunts and horses exercising. Painted in much stronger colors and more continuous in design, these borders would at least show from a distance a definite line of color to outline the panels.

Meanwhile mechanical methods were being steadily improved. Zuber had added to the available range of insoluble colors chrome yellow and mineral blue. The number of blocks employed was enormously increased. It was not at all uncommon to cut three, five, even seven hundred separate blocks for one series, and Dufour for the Laffitte-Madèr Cupid and Psyche set had used fifteen hundred.

In some later landscape papers the number went up close to five thousand.

In 1823 a German, Sprölin, who had been manufacturing wallpapers in Vienna, discovered a way to print rainbow papers, the descendants of the Seventeenth Century marble papers and to become the progenitor of the modern blends. The method was taken up by a cotton printing firm of Ausburg, Schöppler and Hartmann, who thereafter produced these papers.

In 1830 for the first time cylinders were substituted regularly for blocks in printing continuous repeating designs, an idea adopted from cotton and linen printing where it had been in use since 1797, Oberkampf having introduced it at that time in printing his *toiles de Jouy.* In 1835 Bumstead, in England, invented a one-color printing machine. In 1839 this was improved to print four colors. The cylinders for this were cut in relief. The machine was run by hand but nevertheless it could print two hundred rolls a day, an enormous advance over previous production. It was almost immediately introduced into France by Isidore Leroy. Meanwhile in France in

1838 Bissonet had invented another color printing machine. Potter, an Englishman, then perfected a machine similar to the cotton printing machines. In 1844 America got its first color printing machine shipped from England to the Howell factory. By this time the number of colors that could be printed had been multiplied to fifty-four. By 1854 the machine was completely perfected and in general use everywhere. After this wood blocks and hand printing were used only for special designs of some unusual importance.

With the perfection of machinery the making of wallpaper developed rapidly into a highly commercialized business intent primarily on improved methods and quantity production. By 1867 the last of the great scenic papers had been produced. Textile and conventional designs were again in vogue. In the next twenty years under the commercial system and the general deterioration in taste the artistic value of wallpaper declined seriously. Poor colors added to the horrors of bad design to produce hundreds of patterns that had no merit whatever to recommend them.

During this period there was introduced an interesting new type of paper. In 1856 when

Courtesy E. Weyhe

A PAGE FROM ROWLANDSON'S BORDERS FOR ROOMS AND SCREENS.

Japan was opened to Western commerce there was found there a type of paper for walls and screens that simulated embossed leather. The paper itself was of very tough quality, made from the fibers of the plant *Edgemonthis papyrifera.* Three or four sheets of this were stuck together to make one thickness, giving a very heavy, spongy material. The embossing was done with wooden rollers, the printing being under the direction of the Bank Note Printing Office of the Government. After being embossed thin sheets of silver or gold were laid over the paper. The design was then stenciled on and, to complete it, the whole surface was then lacquered. The result was a very rich and handsome paper that found a ready market in the West.

The commercializing of wallpaper design and production provoked one notable reaction. William Morris, at war with machinery and the degradation of the decorative arts that it had caused, established in 1861 the firm of Morris, Marshall, Faulkner and Company at Number Eight Red Lion Square, London, for the design and execution of Mural Decorations, whether pictures, pattern work or only color arrangement, carving, stained glass, metal work including jewelry, furniture, embroidery, stamped leather and so on. In a year the firm added to this list of products chintzes, wallpapers, and carpets. It was in November, 1862, that they attempted their first paper, the rose trellis design, for which Morris himself drew the trellis and vine while Webb did the birds. Hitherto the firm of Morris had itself done all its own manufacturing and at first they undertook to continue the policy with wallpapers. They etched the pattern on a zinc plate and tried printing with oil colors. But the method was found to be slow and unsatisfactory so they substituted instead the traditional wood block printing with tempera colors. In this, too, however, they proved not to have the best facilities, so in the end they consigned the production of their wallpaper designs to an already established firm, Jeffreys and Company.

The first Morris pattern on the market was the daisy design. Thereafter there followed quite a series culminating in the pomegranate. Then Morris turned to other interests, but after a few years he again became interested in wallpapers and produced quite a large second group of patterns. In all, Morris made, in whole or in part, between seventy and eighty wallpaper designs or, counting provisions for different color combinations in the printing, fully twice that number. All of these Morris himself supervised most carefully from the moment the original drawing left his hands until the finished rolls were ready to market. When the block cutter had traced his design it was always returned to Morris to retouch and he maintained constant supervision of the dye pots of his manufacturer.

Beside producing designs himself Morris stimulated other artists to enter the field, the most notable being Walter Crane, who has produced a number of patterns rather in the Morris manner. On the continent, too, his influence was felt, in Belgium Henri van der Velde producing some interesting designs, and in Germany Otto Eckmann and Walter Leistikow.

The Morris-Ruskin revival of art in England was reflected again in another charming series of designs for wallpapers. Kate Greenaway, painter for and of children, published a series of Almanacs of her delightful water colors beginning with the year 1883 and continuing until 1897 with the exception of the year 1896. In 1893 the Almanac drawings of that year were sold to David Walker of Middleton to be used for wallpapers. They were adapted to nursery papers that are still in use, the most charming child papers that have yet appeared.

In recent years abroad there has been some effort to maintain a small proportion of designs of real merit to supplement the regular commercial production. Designers of ability and note have contributed to these patterns. In England notable work has been done by Heywood Summer, Lewis F. Day, Allen F. Vigers, C. F. A. Vaysey, G. C. Harte, H. W. Bentley and J. D. Sedding. In France, Daragnes, Jacques Klein, Scherdal, Camus, René Crevel and Maurice Testard have shown interesting designs at the Salons. America has still to learn the importance of developing and supporting her own school of artist-designers.

CHAPTER II

WALLPAPERS IN EARLY AMERICAN HOMES

FOR forty or fifty years after the first colonists came to America life was too rigorous and the struggle for existence too absorbing for them to give much heed to arts or decoration. Here and there an attempt was made to build some adaptation of an English manor house. Indian Hill, in Newbury, Massachusetts, was built about 1714; the Fairbanks house in Dedham dates from 1636; and the Philip English house of Salem, unfortunately destroyed in 1833, was built in 1685. But these architecturally more self-conscious attempts were an exception. For the mass of the people comfort and cleanliness were the most to which they could aspire. When they had any household treasures they were usually pieces brought with them from their English homes. There was little importation of additional luxuries until toward the close of the Seventeenth Century.

The interiors of the first American homes were, therefore, very simple. Usually the walls were whitewashed but sometimes they were painted with a mixture of clay and water that gave them either a grayish or a yellowish tone according to the kind of clay immediately available. There seems to have been in the beginning no attempt at design or ornamentation.

But the need for decoration soon made itself felt. Walls began to be painted with gayer colors and sometimes there was added a little border across the top or even top and bottom, with a running motive down the side, marking each wall into a panel. These were painted by amateurs, often some member of the family who had a little flair for drawing. They were consequently simple, either the less evolved geometrical forms repeated without any plan or composition, or, in other cases, simple, friendly flowers, daisies, violets, pinks. But some of these amateur decorators were more adventurous. Especially on the chimney breasts they were wont to extend their efforts, attempting a landscape or a romantic scene. And in occasional instances these more ambitious decorations were carried out for whole rooms, either in the simpler patterns such as a repeated running vine, or even with a sequence of scenes and episodes, making quite an elaborate mural treatment.

Wallpapers did not begin to come into the country until the second quarter of the Eighteenth Century. Even then they were specially ordered by the individual householders from their dealers in London or Paris rather than carried for the general retail trade. Thomas Hancock, for instance, ordered in 1737 a specially made paper from his stationer, Thomas Rowe, of London, and in his letter giving the order speaks of another paper similarly imported by a friend of his three or four years earlier. By 1745, however, wallpaper was in retail stock here, Charles Hargrave of Philadelphia advertising it in that year. But certainly it was not in general use until 1750. By that time it had become enormously popular and a host of merchants in all the larger towns were advertising new importations with the arrival of every ship.

In 1765 John Rugar began manufacturing in New York. The next factory does not appear until the Eighties, when John Walsh was operating a plant in Boston. In a few years he went bankrupt and was brought out by Moses

Grant. At the same time Joseph Hovey was printing in Boston both papers and linens and a little later the firm of Prentis and May entered the business, soon, however to split into the separate establishments of William May and Appleton Prentis.

Shortly after this the manufacture was attempted by one Peter Fleeson of Philadelphia, but his firm has left no traces and he seems not to have had any local competitor until 1789, when Plunkett Heeson of the same city undertook a factory.* The next year the industry began here in earnest. The first important firm was founded by two Frenchmen, Boulu and Charden in association with John Carnes, who had been American consul in Lyons. William Poyntell went into the business shortly after that and still a third factory was founded by John Howell and his son John B. Howell at Albany, N. Y. The Howells had good practical experience behind them, for they had been in the business in England before coming to America. They started on a very modest scale, using

* Could Fleeson and Heeson be one and the same name? The confusion can easily be made when the names are written in a document.

AN EARLY NINETEENTH CENTURY FRENCH PAPER SHOWS SCENES FROM THE BOSPORUS.
IN AN OLD HOUSE AT MARBLEHEAD, MASS.

THE PAPER SHOWING SCENES OF PARIS BROUGHT TOGETHER ALL THE FAMOUS BUILDINGS OF THE CITY. IN AN OLD HOUSE AT SALEM, MASS.

as a factory a few rooms in their house, and for some reason soon sold out and moved to New York City, from there to Baltimore and thence to Philadelphia. But the business weathered these vicissitudes and soon became so firmly established that it still persists to-day in the hands of their descendants. The original Howell Factory in Albany was meanwhile carried on for a time by a man named Lemuel Steel. In 1800 Josiah Bumstead began manufacturing in Boston and in 1810 another factory appears under the ownership of Mr. Boriken.

For the really fine papers, however, America still looked to France and England. After the post-Revolutionary depression had passed, wealth had begun to increase and with it the appetite for expensive decorations and house furnishings. Good wallpaper sold at high prices, one room costing at least a hundred dollars with an additional forty for hanging, a really large amount considering the greater purchasing power at that time. This grade of paper America had not yet begun to produce.

American papers in the middle of the Eighteenth Century ran in price from two and five shillings a roll or for pictorial sets as high as sixty shillings. By the close of the century some papers as high as eighteen shillings a roll were being manufactured. Imported papers came in both cheaper and more expensive grades, ranging in price all the way from four shillings for two quires to sixteen pounds for ten rolls.

The taste in wallpapers in the early periods of America ran to strong and varied patterns. Even the floral papers chosen seem generally to have been the more elaborate kinds. Thomas Hancock, for instance, ordered one with a "Great Variety of Different Sorts of Birds, Peacocks, Macoys and Squirril, Monkys Fruit, Flowers etc.," and not content with that suggests that the designer add "more Birds flying here and there, with Some Landskips at the Bottom." He specifies a little enviously that his friend Mr. Waldon had an even greater variety than this on his paper. And the earliest extant example still on the wall of an early American house that is left to us—that in an Eighteenth Century house at Hampton Falls, New Hampshire—shows a hunting scene with the deer and the dog and the hunter with his horn in a three-tiered pattern.

Even the locally printed papers were of ambitious design. Boston manufacturers advertised patterns of Quaker figures, of Diana, of War, Peace, Music, Love, and Rural Scenes, as well as the more usual chintzes, panels and flocks.

Perhaps the reason for the popularity of elaborate and illustrative wallpapers lay in the rather serious limitations of the life at that time. Recreations were few and simple in character, and America was quite isolated from the rest of the world. Travel for amusement was almost out of the question. It was a matter of several days from city to city and a trip across the Atlantic was an arduous and dangerous undertaking, a month's discomfort under the most favorable sailing conditions, and not infrequently six weeks or even two months. Moreover, people at that time did not have as compensation any such vicarious experiences as those offered to us by the illustrated news supplements and moving pictures. Even books were not very plentiful, for there were not convenient circulating libraries to which to resort.

The amusing and imaginative wallpapers could help make up for the rather barren and repetitious days. If it was travel that you yearned for you could take long journeys on your sitting-room walls. In truth, by adding together the various wallpapers one could have taken a trip around the entire world. Going first to Scotland you could have enjoyed the Scottish Scenes, from there to the Lowlands to see the Canals of Holland, then on to Paris, and on one paper have a drive along the Boulevard with its multiple incident, and on another take a tour of the famous buildings of Paris, see the Madeleine, the Trocadero and the other important architectural monuments. In Italy you could make quite a complete wallpaper exploration, see Rome with her most famous ruins, Venice with her canals and romance, and the Bay of Naples with the volcano smoking dramatically in the background. After that, if you chose, you could take a side trip to Spain to see the Alhambra, or, even better, in another pattern mingle with the most romantic Spanish gallants in a serenading, castellated world. Meanwhile there were any number of rural scenes of different types in which you could loiter if you chose. From Europe you could plunge into the Orient, to the Near East first to see Gallipoli and the

Bosporus, get the nice foreign flavor of palms and camels and mosques silhouetted against the sky, then go to see the caparisoned elephants of India and be present at a royal procession in a land that combined the reputed glories of both India and China. If you chose to penetrate further into the tropics you could reach a country never before or since revealed to the tourist eye, a jungle-like forest in which familiar trees were mingled with luxuriant palms, inhabited by savages, some of them dressed like bal masque Indians, others quite early Roman in their habiliments and given to classical games of strength.

If, instead of the tropics, you had chosen to wander more extensively in China, the United States would not have had as wide a range of these travel wallpapers to show you as Europe. The vogue that raged through France and England for the Chinese papers was only lightly reflected in America. By the time America was ready to buy papers production in France and England was well under way, and as most of the importing came via Europe rather than direct from the Orient, European papers took precedence. There were, however, a few of the real Chinese papers. One could, for instance, see the whole process of the manufacture of tea on one.

To complete the tour around the world you would go next to Peru and have as companion no less a person of distinction than Pizarro himself. Finally you could end in America and see the sights of your own country on the four sides of your room.

All of these papers were not absolutely contemporaneous, but they were all made during the end of the Eighteenth Century and the beginning of the Nineteenth and were all to be seen on the walls of at least one and in some cases a number of the mansions of early America. Most of them were in the northern states, New England and the vicinity of New York. In the southern mansions preference ran much more consistently to plain paneled walls, or, if paper was used, an architectural design or a textile pattern was most frequently employed. There is, for instance, an interesting paper showing a repeated bas-relief of a chariot that was found on the walls of an old house in New Orleans and has been commercially reproduced. Only occasionally are the picture papers seen in the early southern homes.

If in the early days in New England or New York your taste had gone rather to literature than to travel, wallpaper would have offered entertainment here, too. You could have had a sort of decorative, arrested moving picture of your favorite story on your parlor walls. There was here, too, quite a range from which to choose. The Bible stories were given quite completely. You could see in one set the Patriarchs, Abraham offering up Isaac, Joseph with his coat of many colors thrown into the pit by his brethern, Noah's Ark, Pharaoh and his Host in the Red Sea, Rebecca at the Well, and Moses in the Bullrushes. If your interest leaned rather to the classical you could have all the gods and goddesses of Greek mythology, Aphrodite, Jove, Hermes, Hephaestus and a host of others great and small. Or if you wished more incident, you could have the tale of Telemachus on the Isle of Calypso, an apochryphal appendix to the Odyssey invented by Fénelon and very popular in this country for wallpaper. Or, even finer in design and execution, there was the story of Cupid and Psyche designed by Lafitte, the cartoons painted by Madèr père and published by Dufour. For the more romantic soul there was Robinson

Crusoe, Captain Cook's Adventures, very realistic and cannibalistic, Don Quixote, or, later, the Lady of the Lake.

For the potential devotees of the rotogravure sections there were wallpapers of current or recent historical events. So you could see all of Napoleon's campaign in Egypt, a beautiful and richly colored paper. On another paper Cornwallis presented his sword to Washington with a perpetual and reiterated gesture of defeat, a valuable bit of propaganda for promoting national pride and patriotism. Still other papers showed horse racing, hunting scenes, or the daily life of the French soldier.

Most of these pictorial papers were on a very large scale, the figures two feet high or more and the panoramas so extensive that one scene filled an entire wall. Some of them, however, were done in small scenes, each surrounded with a formal scroll or floral pattern. The Alhambra, for instance, was done in a small repeating pattern and some of the hunting scenes were part of a verdure decoration.

Among the smaller scale papers were the more conventional designs, too. They were dictated, of course, by the prevailing fashions of France, but America showed her individ-

Photo by M. E. Hewitt

THE SCENES OF AMERICA INCLUDED BOSTON HARBOR. A MODERN REPRINTING OF AN EARLY NINETEENTH CENTURY FRENCH PAPER.

Photo by M. E. Hewitt

CONVENTIONAL BLOCK PATTERNED PAPERS WERE FREQUENTLY USED IN EARLY AMERICAN HOUSES. A MODERN REPRINTING OF AN OLD DESIGN.

uality by selecting certain types of patterns. The more elaborately delicate Louis XVI patterns in brocade or Pompadour or moiré effects were little used, as were the more extravagant of the Empire and Directoire periods. America picked out floral designs, some of the *toile de Jouy* types, with small scenes in grisaille at repeated intervals in stripes, and occasionally the two-toned stripes. With the advent of Chippendale came some Chinoiseries, but Chinese Chippendale was not as popular here as the more conservative Chippendale designs. As the Nineteenth Century advanced, floral papers came more and more to the fore, the wallpapers being a counterpart of the prevailing modes in chintz, with large luxurious blossoms, peonies, large, full, bright-colored roses, or abundantly bearing grapevines, often with brilliantly plumed birds perched on them. The large scale textile patterns came back into their own, too, with a preference for heavy damasks and velvet effects.

The early Americans were very proud of their gorgeous wallpapers, and justly. There is an amusing legend of Martha Washington in which it is said that the good housewife, when she was about to entertain General Lafayette, was greatly exercised for fear that the new wallpaper she had ordered to furnish the house for the occasion would not get hung in time. The paper finally arrived after Lafayette was already there. He, seeing her anxiety about it, offered to cooperate with General Washington in putting it up. So the two generals turned paper-hangers.

There is at least this much to substantiate the legend, a note in Washington's own handwriting carefully recording the "Upholsterer's Directions" that whitewashed walls require glue whereas otherwise simple paste would do, "but, in either case, the Paste must be made of the finest and best flour, and free from lumps. The Paste is to be made thick and may be thinned by putting water in it. The Paste is to be put upon the paper and suffered to remain about five minutes, to soak in before it is put up, until all parts stick. If there be wrinkles anywhere, put a large piece of paper thereon and then rub them out with cloth as before mentioned." Certainly the Father of Our Country had at some time had occasion to have a practical interest in wallpaper hanging.

Franklin, too, took a personal interest in the papers for his walls. Writing to his wife from London in 1765 he suggested: "Paint the wainscot a dead white, paper the walls blue and tack the gilt border round the cornice. If the paper is not equally colored when pasted on let it be brushed over again with the same color and let the papier-mâché musical figures be tacked to the middle of the ceiling. When this is done I think it will look very well." The consideration of having to paint the paper after it was up refers to a letter of his wife two years earlier, in which she had complained that in the Blue Room, evidently just done over, "the paper of the room has lost much of its bloom by pasting up." The dyes were not yet absolutely insoluble.

With the revival of interest in all things early American these papers that early America imported from France have come back into their own. Where the originals remain they are much prized, but originals are scarce. There has been, therefore, a campaign to recover the original blocks from which these papers and others like them were printed, and new printings are to-day being made. Where even the blocks have been lost, careful reproductions of some patterns have been made, either in the original wood block techinque or, where it is possible, in metal rollers for mechanical production.

Certainly no early American house, especially in the New England style, whether it be old or modern copy, is perfect without some of the old-fashioned wallpapers. They complete the quaint and straightforward attraction of this style of architecture, perfect the reproduction not only in historical accuracy but in spirit and character. They combine a naïve interest with a sophisticated sense of decoration to make a wall finish that has at once individuality and wistful charm.

CHAPTER III

THE MODERNIST WALLPAPER

THE reaction from machine-made decorations and from their neglect of design and the stupid, repetitious similarity which had begun with Ruskin and Morris, culminated at the end of the Nineteenth and the beginning of the Twentieth Centuries in the Arts and Crafts movement and the development of what came to be called the Art Nouveau. The trend was expressed in different ways in different countries, but all phases of the movement had in common a self-conscious care to revive design, an emphasis on the technique of production that made itself decisively felt in design, and a complacent awareness of responsibility for founding a modern style that would take its place in the historical succession.

In America the tendency was felt in the transient vogue of Craftsman furniture, with the accompanying accessories of hand-beaten metals, hand-woven materials and hand-tooled leathers. In Europe the result was a more ambitious type of decoration, more lasting in its influence. The movement was strongest in Munich with a closely related school in Vienna. The outstanding products were a rather pretentious and intentionally artistic kind of furniture, conspicuously and deliberately different from all commercial products, simple and massive, with heavy arabesques and unusual silhouettes, and a full equipment of all minor arts to complete a room scheme. In color these decorations departed violently from the muddy colors, limited range of tones and random combinations of the usual market wares, using, instead, vivid hues, the less usual shades such as burnt oranges and peacock blues, and artificially contrived and rigidly adhered to color schemes. The color schemes were considerably influenced by Japanese prints which had been brought to the fore by Whistler and the Impressionists.

The Munich School of Art Nouveau has produced numbers of wallpaper designs in accordance with its own ideals. They are usually heavy in outline to accentuate the printing method, simplified in drawing to get away from any taint of naturalism and done in two or three bright and sharply contrasted colors, preferably the less common shades of orange, purple, green, and blue. Notable among the designers are Emmy Seyfried, Adelbert Niemeyer, Otto Gussmann, W. V. Beckerath, Rudolph Rochga, Carl Weigl, Paul Haustein, Josef Hoffman, Ernest Riegal, Karl Brauer, Lotte Frömmel and W. Zurmühlen.

In France the Art Nouveau movement has been led by such really notable decorators as Gallé, Grasset, Dufrêne and Follot. These men have been feeling around for a genuine and individual style that shall be not only modern but unmistakably French. Because they are men of taste and ability they have produced some interesting and delightful furniture and a few attractive experiments in the decorative arts. But their work has been, not really French, but derivative. At its inception it had an unwholesome reverence for the Munich School which was older, better established and more prolific, and later it relied too often on the relics from the past in the museums, adapting or even copying these. With these it combined a bastard Orientalism compounded of, first, the Japanese influence already operative in the Munich work, second, the suggestions derived from the costume designs and stage setting of the Russian ballet that so captured the Paris imagination, and, third, of a theatricalized understanding of Persian art which was rediscovered with great acclaim in France fifteen or twenty years ago.

An art so heterogeneous in its materials could not, of course, found a national style. Nor could it, as a matter of fact, become a true style at all. For it was the work, not of a people building on its past to fill its guiding needs, but of a few artists working without background, without contact with their public, even without cooperation among themselves and without any clearly defined set of ideals or program. The most that these men could produce under these circumstances were a few designs doomed to remain aloof from the historical trend as individual eccentricities without any possible issue.

About 1909 a group of younger men got together seriously to undertake to build up a modern French style. These men profited by the mistakes of their predecessors by avoiding them. In the first place, they sought contact with the people and the life of France, both in their historical traditions and deeply founded

Courtesy Henry Bosch and Company

A MODERN GERMAN DESIGN THAT HAS FELT THE INFLUENCE OF CUBISM.

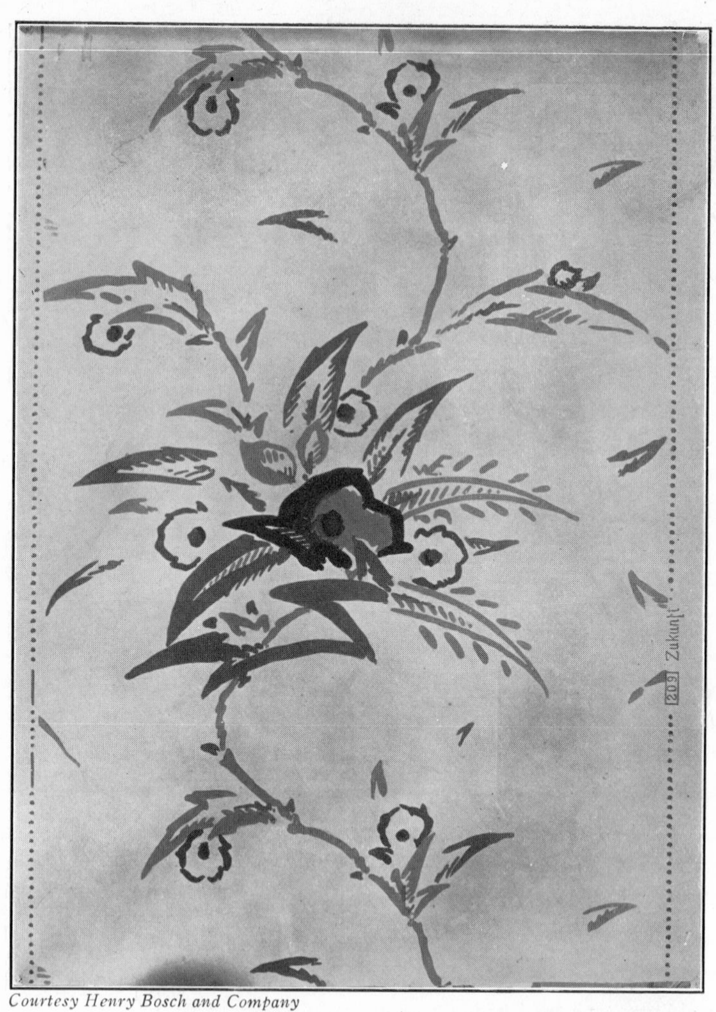

Courtesy Henry Bosch and Company

A MODERNIST PAPER FROM GERMANY.

habits and tastes and in their current needs and interests. And, in the second place, they formulated a set of ideals and a program for the accomplishment of which they agreed to cooperate to the full, setting aside personal fame and gain for the better development of a modern, national style.

To establish sound connections with French life this group turned back to the Provinces where traditions hold and the conservative and deeply French middle class lives a life that is continuous with the life of their forebears for generations, and so is profoundly national. Here, among the families of this class, they found still in use the furniture of Louis Philippe and Napoleon. This style they took for the starting-point of their design, first because it was actually in use by the most representative section of France and so would insure continuity with the national spirit, second because it had proved itself congenial to that spirit by its long endurance in the people's affections; and third because it does fulfill the recognizable common character of the French, their logical common sense. For these two closely related styles of design are built of lines and forms stripped to the bare essentials and composed in clear, unmistakable, necessary relations, so that they bespeak reserve and restraint. They are as congenial to the hard-headed, exact-minded French bourgeoisie as their clear, perfect language.

So the decoration of Louis Philippe and Napoleon was one foundation stone that these younger men found for the modern French style. The other seems, at first glance, in capricious contradiction. It is the modern school of painting that issues from Cézanne and has taken one conspicuous form in Cubism. But eccentric as the combination of provincial Louis Philippe and the modern art may seem, for the purposes of this school it is really rational, for the style for which this group is aiming is not merely French but is also modern. It must express not only the old French character, but, too, that same character as it has been shaped by modern life. And the modernist painting is the spontaneous expression of modern life. Nor is it really in such sharp contradiction to the Louis Philippe of the bourgeoisie. It, too, seeks the essential forms stripped bare of the trivial, the accidental, the superfluous, seeks these forms in their most significant revelations. And too, it looks for the absolutely necessary relations. The rather arid furniture of Louis Philippe and the apparently bizarre creations of the new art are two attempts to formulate the work of strictly schooled rationality.

The men who got together in 1909 to compose this group had already been working singly, or in informal combinations, at modern furniture and decoration designs, and been exhibiting at the Salons alongside the older Art Nouveau exponents. There was André Mare, head and creator of the association, painter and designer; Duchamps-Villon, architect and sculptor whose production, that already showed greatness, was cut short by his untimely death during the War; La Fresnaye, Richard Desvallières, André Versan, Richemont-Déssaignes, Jacques Villon, Paul Véra, J.-L. Gampart, Marie Laurencin, Madame Lanoa and Madame Sabine-Desvallières. Perhaps the most significant thing about the undertaking was the fact that these men were not merely decorators and designers following the lead of the new school of painting and adapting it at second hand to the lesser arts, but they were themselves the creators of that school, the makers of the art that they were applying.

For the first time since the Renaissance the men who were painting the pictures and modeling the statues of the prevailing school of art were also designing the houses into which those paintings and statues were to go, fashioning the furniture, dictating the fabrics and the incidental objects. In the Sixteenth Century in Italy great painters had been architects and engineers and goldsmiths. In the preceding centuries in France the relation had been reversed, and the craftsmen had painted the pictures and carved the statues. Either way the arts keep healthy by a normal continuity. But since then the so-called fine arts, becoming more and more snobbish, have drawn ever further away from the lesser branches, leaving them to wither and die. Now in the modernist group the natural cohesion was reestablished and each could nourish the other.

In 1912 the group undertook to build and furnish an entire house as an example of their principles. In this case André Mare and J-L. Gampert designed the wallpapers, but one of the characteristic features in their organization was that their functions were interchangeable. The furniture designer to-day might be making book covers to-morrow and the wallpaper designer have turned back to architecture. Besides the designs by Mare and Gampert, wallpaper patterns have also been made by Marie Laurencin, M. Constance Lloyd, an English woman who was associated with them while residing in Paris, Albert André, G. d'Espagnat, Laprade, Lebasque, Barbier, Carlègle, Drèsa, Herman-Paul, Süé, Paul Véra and André Groult, who undertook to supervise the printing for all of them.

For the most part these men have returned to nature for the elements of their designs. Groult has done one or two papers using the traditional Louis Philippe motives but usually the Louis Philippe influence has made itself felt in the papers only in the closely maintained coherence of all parts of the design, their rigid rationality and careful adjustment in proportion to the spaces to be filled. The subject matter has come from nature as seen and interpreted by modern painting.

Flowers, singly, in bouquets, and in baskets, are most frequently used, but fruits too have been used, either by themselves or in conjunction with flowers, and birds appear rather often. The human figure they have only occasionally adapted. Paul Véra, for instance, showed at the Salon des Artists Décorateurs in 1920 a wallpaper pattern with an all-over design of a rose-bush, thick with foliage and blossoms, with birds perched on it, that broke away at irregular intervals to reveal four different groups of nudes. But this use of the human figure has been the exception rather than the rule in the modernist patterns.

Their rendering of flowers and fruits is quite different from any hitherto used in wallpaper design. It is, of course, utterly remote both from the gross naturalism of so many current commercial papers, with their heavily modeled roses and birds of detailed plumage, and from, too, the more delicate exactness of the old Chinese papers with their meticulous botanical and ornithological portrayals rendered in flat silhouette. But it is also equally remote from the conventionalization, the imposing on the flowers of an arbitrary geometric form, of the late Nineteenth Century, and from the impressionism of the early Twentieth, which used flowers only as an excuse for blobs and splotches of color. These modern men have gone back to the flower itself, looked at it with respect, reverence almost, without the distorting prepossession of any traditional rendering, and then have tried to analyze out of it its most typical, essential and significant form. They have proceeded in designing just as they have in drawing their still-life studies.

The effort to get the essential form has, of course, led to simplification. In the work of different men this simplification has taken different forms. In the paper of roses and nudes of Paul Véra, for instance, the nudes are portrayed in outline in long, uncompounded curves that give a silhouette rather than a detailed drawing of the human body, a synthetic impression of it. The flowers, on the other hand, are geometricalized, but it is not the old-fashioned kind of geometricalizing that imposes arbitrarily a formal outline on a flower, but the new kind of geometricalizing related to Cubism that starts with the actual shape of the flower, then gives it form by prolonging certain lines and emphasizing others, or dissociating certain sets of lines that have a vivid interrelation, and setting them forth to represent the flower. In certain other modernist designers, such as André Mare, the geometrical form evolved from the flower is chosen in such a way as to make clear the structure or architecture of the plant. But in every case the flower still retains its individuality and, especially, its vitality.

The simplified forms of the modern school lend themselves excellently to wallpaper design. They have, in the first place, the strong silhouette character that is necessary, in the second place they are sufficiently clean in outline and mass to give an impression from a distance, in the third place they are primarily flattened forms, and in the fourth place they are drawn forms rather than painted forms, and so can be appropriately printed.

The modernists have completed these well-adapted forms with clean, vibrant, varied colors. Vigorous polychromy is, in fact, so characteristic of the school that when in 1909 Süé, Huillard, Mare and Groult first began to formulate the style they were known as "colorists." But their color, though quite as strong as that of Art Nouveau designs, differs from it by turning away from the queer and unusual shades to the clean, strong tones of the familiar primary colors, and using them, not in fixed and obvious relations, but in lavish variety.

The bigger men and the originators have inevitably inspired imitators and it is a fatally easy style to imitate. So there have been a number of modernist and Cubist papers trivial in style and really meaningless. But some fine designs have been and are still being produced both from the French School and from the very active modernist groups in Germany, and the unavoidable poor examples constitute no real criticism of the style, a style that offers wallpaper designers an opportunity to invest their art with vitality again by bringing it back into touch with the current art and life.

CHAPTER IV

THE MANUFACTURE OF WALLPAPER

THE mechanical processes of manufacturing wallpaper are very simple. The usual factory has only three kinds of machines with, perhaps, one or two variants of one of them, and these machines themselves are very simple. A few factories prepare their own raw materials—make, that is, their own paper and their own dyes, and in this case there is added to the wallpaper plant itself the more complicated plant for the paper making, and the laboratories for the dyes; but the ordinary factory buys both paper and dyes, and undertakes only to print, emboss and prepare for market their product.

All mechanical wallpaper printing is done on a cylinder press. In this there is a large central drum of metal that is covered with a lining of canvas or, for better work, thick rubber to make a slightly soft printing surface. The paper in a huge roll of 1500 feet passes off a reel over several small wooden rollers that serve to keep the tension even, down under this drum. As the paper passes down under the drum and around it, a series of cylinders roll over its surface and print it. Each of these cylinders is engraved in relief with one part of the pattern which corresponds to one color, and they print in succession in such a way that the registration falls exactly side by side with no overlapping. Each cylinder is continuously inked with its own color out of a trough of the liquid dye which is below the cylinder. In this trough a small roller constantly revolves to keep the color mixed, and a continuous porous cloth, called the sieve cloth, which is stretched over two or three rollers, passes through it unceasingly, carrying the color up to the printing cylinder. Thus the printing cylinder is in contact with the always freshly moistened cloth on the one side, taking the color from it, and with the steadily moving paper on the other side, applying the color to that. The number of different cylinders that can register in succession, and so the number of colors that can be printed at one run, is theoretically unlimited, but actually twelve is usually the largest number of cylinders used, though occasionally there are machines employed with as many as eighteen cylinders.

If the colors of a design are all to be printed side by side, the entire pattern can be printed at one run of the paper through the machine. If, however, one color is being printed on top of another, as in a colored design on a colored ground, the first colors must dry before the second can be put on. This drying is done in a special machine, which also runs automatically. The paper passes off the printing machine and falls over a bar run on a carrier composed of two horizontal beams just below the ceiling of the factory. The bar moves on, and the paper is caught again over a second bar far enough from the first so that when the two run together, or within about a foot and a half of each other, the paper makes a long loop almost to the floor and back. This is repeated with the whole series of bars until the paper is draped in a succession of deep loops from the ceiling almost to the floor. The whole series of loops then moves slowly between the hot boxes, compartments about twelve or fifteen feet long and the full height of the room, filled with steam pipes.

In the old type of drier the drying depended

entirely on the intense heat from these steam pipes. As a result, a wallpaper factory was a most unpleasant place in which to work, very hot, and with the atmosphere saturated with the heavy steam from the drying papers. This moisture of the atmosphere also prevented the paper from drying as quickly and as thoroughly as it should. To correct these evils, a modified drying machine has recently been perfected that depends, not primarily on heat, but on fresh air. In this, the paper looped on the rollers passes between the hot boxes in the same way, but the steam coils need to be kept hot only when it is necessary to dry the paper very rapidly. For in addition to the coils there is a large electric fan forcing fresh air over the paper constantly, and, above, an exhaust pump removing the steam-saturated air that comes off of the paper. So the paper is dried without so much heat, is dried more thoroughly, and the atmosphere in the factory is kept clean and fresh.

As the paper comes off the drying racks it is automatically rolled into a large roll. If it is a design that cannot be printed at one run it must be put through the printing machine again to get whatever additional colors are needed.

If the paper is to be only a printed paper it is then ready to be rolled into the market lengths, labeled, and the job is done. If it is to be embossed, as many papers to-day are, it must go through some kind of embossing machine. This may be to emboss the design in relief or to emboss the whole paper into some kind of rough texture. In either case the paper is run through a press between two rolls the upper one of which has the design to be embossed in linear relief, and the lower one of which is made of a rather soft material, usually a form of paper composition, so that the paper between the two will receive a deep impression. If the design is to be embossed in relief the upper roller has the outline of that design raised in metal wire, and the paper is accurately adjusted so that the outline on the cylinder will coincide with the already printed outline on the paper. If it is a general texture that is to be obtained, the cylinder has on it a series of brass ridges that vary in design according to whether the desired effect is a linen finish, a burlap texture, a grass cloth or any one of a number of woven qualities.

There are occasional papers that depend entirely on the embossing for their design. In these, the relief design is usually very heavy, and so they require a heavier quality of paper. Such a relief design may be worked out in plain white paper, in which case a small amount of metallic color is sometimes stamped into the background with the embossing machines, so that the product is finished in one process; or the relief may be employed on a one-color paper, the embossing catching the light in such a way as to make it look two-toned. This latter type of pattern is especially effective on a metal paper where the light is sharply reflected from the relief, so that there is a decided contrast in the two tones.

The cylinders for printing colors may be all metal or metal and felt. The all-metal rollers are used, for instance, in jaspé or strié backgrounds. The cylinders for these have a series of broken lines attached, and so arranged that the succession of colors is printed immediately side by side in narrow lines or small splotches, the lengths and areas of the juxtaposed colors varying irregularly so that when finished the effect is not of a line pattern but of a play of mottled colors.

The imitation grass cloths are similarly printed and then usually embossed. The designs for some of these paper grass cloths have been obtained by photographing a real grass cloth and making the printing cylinders directly from the photographs, so that the paper is actually a printed facsimile of the real fabric.

Cylinders entirely of metal are sometimes used, too, for printing the usual patterns. On these cylinders the design is cast in relief out of type metal, just as regular type is cast. This is a very inexpensive type of cylinder to make, but it is not inexpensive to use for regular production, because the type metal is so soft the edges of the pattern soon get blurred, and the cylinder has to be remade. Ordinarily, type metal cylinders are used only for a special design of which a very small amount is needed, or for an experimental run when a pattern is first being put on the market to see if it is going to be a sufficient success commercially to warrant continued production.

On the usual color printing roller the design is made of a brass outline filled in with felt. The brass wire, about a sixteenth of an inch high, makes a little, shallow well, as it were, into which the felt in a pasty state is compressed. This felt dries very hard, making a solid yet slightly resilient surface which gives the best

Courtesy of Robert Graves and Company

MAKING THE ROLLS FROM WHICH WALLPAPER IS PRINTED.

Courtesy of Robert Graves and Company

THE TWELVE-COLOR PRINTING MACHINE AT WORK.

printing, while the brass outline keeps the edges clean and clear. While such a roller as this costs about three hundred dollars to make and the type metal costs only about a third as much, the more expensive roller is in the end the cheaper because it never wears out.

The dyes used in the wallpaper printing are almost entirely synthetic water colors. They come by the barrel in a pulpy or pasty form and the manufacturer has only to add hot water and a certain amount of size, or heavy glue, to increase the adhesiveness. Metal colors, also, now come in this form, a comparatively recent improvement that makes the production of papers with metal colors much cheaper and easier than it was two decades or so ago. If the design requires an especially dense or brilliant color a small quantity of aniline dye is sometimes added. A few papers, such as some of the leather effects, employ oil colors, which are more expensive but are also more durable.

The flock papers are made by the same type of machine as the printed papers, but instead of being registered in colors the design is printed in a kind of glue, and while the glue is still wet the paper passes under a box from which the chopped or powdered wool or silk is sifted heavily down onto it, then it passes over another simple machine which shakes it sharply, thus knocking off all the extra flock.

Most papers, whether printed, embossed or flock are made on a white or écru-colored paper. If a colored ground is demanded, this is printed on with a plain roller. But there are a few kinds of paper that are dyed in the pulp. The commonest of these are the oatmeals and the crêpes. Pulp dying usually gives, of course, a one-colored paper but certain pulp-dyed crêpe papers have also a two-toned, blended effect.

In addition to these common processes there are a few special processes used to give a satin finish, a moiré effect, a glazed ground, and so on, processes that have to do with the finish of the paper. There are, too, a few secret processes for making especially heavy flocks or giving certain unusual texture effects. But all these are only variants of the fundamental common methods of printing and embossing.

When the paper is finally finished, and this seldom requires more than three treatments and is often finished in one, it is rolled off the large roll into the regular market lengths, eight yards to a roll or sixteen to a double roll. There are a few papers, too, of the more expensive grades that are sold by the yard. The final reeling can be done automatically but is often done by hand. When the reeling is done by hand, the paper is marked every eight or sixteen yards in the last process to indicate to the operator when to let down the knife to cut off the length. If the last process has been embossing this mark is a small tear by the embossing machines, if printing a little blotch of color.

The final rolls then have the manufacturer's label, which also serves as a wrapper, attached. On this label is noted both the number of the design and the number of the run, for in every new run or printing there is apt to be some slight variation in color so that it is impossible to use some rolls of one run and some of another together in one room.

As to the designs for wallpapers, the American manufacturers get them here, there and everywhere. Few, if any, of the factories maintain their own designers. They count on buying their patterns from outside artists. There is a healthy tendency to use the museums for historical designs, especially for the textile patterns that are so useful and so popular. The manager or owner of a factory, browsing through a gallery, will see a pattern that appeals to him. He calls on some draughtsman who has worked for him before to make a drawing of it, and together they make any adaptations that may be necessary for the mechanical process. The number of designs needed by the wallpaper trade is enormous, for the combination of our commercial system and the fickle and untrained taste of the public operates to create a constant demand for new patterns, so that a few are continued for more than two or three years and many last only one season. The rapid change in designs is exceedingly unfortunate for it means that the manufacturers cannot afford to pay high prices for them, and so no artists of great ability are attracted to the field. So the designs remain for the most part decidedly inferior in originality.

The method of marketing wallpapers, too, operates against its best artistic development, for the public has no opportunity to express directly its preferences and so make any improvement in its taste felt. It must take what the wholesaler and the retailer are willing to offer it, and the wholesaler and the retailer are business men, tight tied with tradition, and constantly haunted by the consciousness of their financial risks. But in spite of these difficulties the taste in wallpaper is constantly improving and there are now on the American market a number of designs of real quality and distinction.

CHAPTER V

THE PROBLEMS OF WALLPAPER DESIGN

IN making the pattern for a wallpaper the artist must never forget that it is to be the decoration of a flat surface that will be part of the structure of the room. From these two points, the flatness or the surface and the structural function of wallpaper, follow the most fundamental limitations and requirements of the design.

Because the surface is flat the design cannot be either in high relief or in deep perspective. Drawing or coloring that will give the effect of heavy modeling contradicts the appearance of a plane surface and at the same time weakens the effect of the structural firmness. Similarly, remote vistas in a landscape paper disturb the necessary straight rigidity of the wall. If the design is in relief, as some flower designs as well as designs of architectural ornament are, the apparent depth of it should not be any greater than the depth of wood carving or that of a bas-relief. These being actually structural materials they are a good guide to the limitations of paper design where the appearance, only, and not the fact of structural integrity must be maintained. If the design is in perspective, the best guide is the tapestry of the Sixteenth and Early Seventeenth Centuries, when perspective was admitted in the cartoons as a necessary feature of the illustrative scenes, but was formally and decoratively treated by being reduced to a fixed number of planes patterned one above the other in quite clearly indicated strata. In this conventionalized perspective diminution of size only is regarded, the vagueness of atmospheric perspective being disregarded except for a heightening of the color key. So each plane is represented as an irregular, horizontal stripe, decorated with clearly drawn trees, flowers, and buildings. Thus both the decorative richness of the textile and the flatness of the surface are maintained.

Again, because of this necessary flatness in effect, the design must avoid any blank spaces that will give the appearance of holes when the paper is in position. Particularly the semi-conventionalized, all-over flower and verdure patterns are in danger of breaking away into empty spots that look like perforations or large moth holes, giving the exceedingly unpleasant effect of a riddled wall, an effect that not only conflicts with the flatness and solidity of the wall, but destroys also that protected feeling of inclosure which we seek in a room. For similar reasons, such designs must be quite evenly distributed, not concentrated at certain points and thinned out all around them. A continuous design, that is, must be perfectly continuous to succeed.

The need for continuity, moreover, makes it imperative that the unit of repeat in an all-over pattern be not too clearly delimited. It must be so contrived that it runs along, without stopping or hesitating, into its repetition. Otherwise the wall will seem to be broken into sections, an appearance running counter to its structural rigidity. Moreover, if the starting and stopping points of the unit of design on all four sides are not concealed, the paper will hold the attention of any one sitting in the room to its persistent reiteration, a most annoying kind of insistence in design, like that of a tune running unbidden in your mind, which destroys the comfort and beauty of the room. So the artist must manage his pattern so that it is coherent not only within itself, but with the next repeti-

ollection of W. R. Hearst, Esq.

PERSPECTIVE MAY BE ADAPTED TO A FLAT SURFACE BY CONVENTIONALIZATION IN HORIZONTAL BANDS.

Courtesy Charles Grimmer and Sons

PLANT MOTIVES EITHER IN THEIR NATURALISTIC DRAWING OR CONVENTIONALIZED INTO A PATTERN ARE A FUNDAMENTAL MATERIAL OF WALLPAPER DESIGN.

tion of itself, also, and, what is even more difficult, with the other repetition of itself that will be beside it when the widths of paper are pasted on the wall. One of the most successful ways of concealing this joining of the width is called drop-repeat in which the unit of design in one strip begins at the center of the units on the strips at either side. When the pattern is not an all-over design, but is made of separate units, such as individual bunches of flowers on a plain background, or on a conventional background, continuity is maintained by spacing the units so that the intervals between them will not be too great. If they are set too far apart the pattern will have the unpleasant appearance of seeming to stop and start again with every repeat.

Structural fidelity makes other demands on wallpaper design. The design, for instance, must seem to cooperate with the rest of the wall in supporting the ceiling. This is the advantage of stripes, or of any arrangement that suggests stripes. The recurrent vertical lines carry out the work of uprights holding the ceiling in place. But even in an all-over pattern this same vertical emphasis must somehow be implied. A continuous design, for instance that broke up into predominantly horizontal lines would be quite impossible.

It was partly a feeling for structural fidelity, and partly an analogy from the world of nature that made Walter Crane insist on the importance of having a wallpaper design grow up from the bottom and spread out laterally as it grows. A design that seems to drip down from the top, such as, for example a wistaria vine in blossom, falling from a trellis, with no evidence of supporting trunk or pergola framework, would be disturbing and hence unattractive, both because it would contradict our usual experience and expectations and because it would have no indication of the construction of the sustaining wall.

Walter Crane has pointed out also this use of structural lines in ceiling papers. The design of a successful ceiling is, from this point of view, difficult, for you have a large flat surface held up only at the sides. Yet if it is going to be pattern at all, it is logical to put decided emphasis at the center where, structurally, it would be too weak to stand the greatest weight. So, in our design, we have to solve this apparent contradiction. The most satisfactory solution, as Walter Crane points out, is to let the framework of the ceiling appear in the pattern, trace with some emphasis the beams and joints and their interrelations, or make your divisions into panels, or radiate your main lines from some definitely indicated structural center. The type of ceiling that Sheraton has designed is also sound, with no ornamentation in the center except possibly a fine strip that does suggest the construction and the weight of design in borders near the edges.

With high relief and deep perspective ruled out, and continuity through the repeats and the emphasis or structural lines insisted upon, we have only the bare skeleton of wallpaper design. With what is the skeleton to be clothed? Primarily with natural forms or with these same forms conventionalized. Leaf shapes, buds and flower patterns, the branching of a vine, these are subjects of textile ornament of all kinds. They may be conventionalized almost beyond recognition, and the conventionalization may have been determined by the translation of the pattern into some particular medium, stone or wood carving or a woven fabric, but behind those superimposed forms there will still be the natural object. These natural objects, with a few simple geometrical shapes, the triangle and the circle and their parts and combinations, and the elaboration of these into such traditional patterns as guilloches, are the available materials for wallpaper design.

Ingenuity has been stretched to the utmost to find new subjects in the world of nature. The simpler flowers have been repeated unnumbered times and in every possible form and combination, roses and daisies and fruit blossoms, from the Seventeenth Century on. The common wild flowers, buttercups and cowslips and marguerites, had their day in the early Nineteenth Century, but they persist still in papers that want to be ingenuously old-fashioned. The grapevine was popular then too, and the peony, which first came into use in the Chinese papers of the Seventeenth Century. Irises and daffodils appear from time to time and in more recent days the more sophisticated hothouse plants have been introduced, especially the delicate and naturally beautifully patterned orchids. At least one ingenious designer has discovered the multiple possibilities in the fungus realm and made patterns of toadstools and the feathery coral mushroom. Trees and their foliage are an inexhaustible standby.

The beasts have been added to the flowers. The birds are the most generally useful. Pheasants are too gorgeous to have escaped an endless reiteration, beginning with the old Chinese painters; and the parrots and their exotic cousins have been in demand; but the less obvious birds have been as popular, the robins and the finches and the humming-birds. The larger animals are a bit more difficult to handle but they are an important part of many of the landscape papers and a recent paper from England has experimented with the French Gothic *mille fleurs* type of pattern, with little animals running through a thickly flowered wood.

Designers have searched, too, through the history of design for inspirations. Almost no type of textile has been overlooked. An interesting paper has used one of the characteristic Byzantine designs with opposed figures in a circle such as are found on the remaining fragments of Tenth and Eleventh Century silks. Another has repeated without elaboration or adaptation a familar heraldic lion. All the types of damasks, velvets and brocades of Europe of the Seventeenth and Eighteenth Centuries have been repeated in paper, both flock and plain printed, at the time of the fabrics themselves and, in modern days, in the machine-made products. Cordova leather has been reproduced both exactly and in adaptations. The verdure tapestries have provided both suggestions and completed models. Cotton and linen prints, both the simpler calicoes and the more elaborate chintzes, have been fruitful sources for two hundred years, sometimes the paper being printed directly from the same blocks as the linen. Tiles of many types have been used. Lacquer work has been simulated in both design and finish. Suggestions have been taken from Chinese porcelains. And, of course, all types of architectural ornament have been called upon, frescoes being especially adaptable, but bas-reliefs, carved moldings and the abstract conventionalized designs that are heritage of this ancient art coming into play also. The scenic papers have found ancestors in mural decoration, tapestry and even in book illustration.

But though the designer can appropriate from almost all the arts, and nature too, he must usually refashion his material to the needs of his particular technique. If he goes direct to nature, even though he eschews conventionalization he cannot be unrestrainedly naturalistic. He must formalize his plants, must not have, as Walter Crane points out, too pictorial or too sketchy a treatment, for this conflicts with the necessary repetition which gives the pattern an unavoidable formal character. And the pictorial or sketchy treatment would be unpleasantly at variance, too, with the mechanical method of production. Moreover, the designer must fit his natural forms into a systematic plan. Usually this plan will take the form of a geometrical substructure, but if it is not so rigid as this, there will at least be a completely thought out and coherent arrangement. Even the individual sprays must be fitted in to a definitely planned form, with an enclosing outline clearly indicated when not actually drawn.

William Morris is particularly insistent on the importance of this underlying structure of the wallpaper design and has worked out with great care the possible variations that can be used for it. There are, he says, three qualities to be sought in any pattern—beauty, imagination and order—and of these the most important is order, for without that neither the beauty nor the imagination is visible. This order can be arranged in any one of four general forms: stripes, checkers, diaper patterns, or the continuous growth of the lines. In a pattern that is striped or suggests stripes the beauty will depend fundamentally on the proportions of the stripes and the interrelations of the colors. If they are undecorated stripes the sheer beauty of the material, the texture effect of satin or moiré, for instance, will be important. A checkered pattern is also a problem in proportion and color. When you get to the more complicated diaper patterns you have a wider range of possibilities and, in consequence, a more generally applicable type of scheme. The diaper may progress in circles or it may be floriated, it may branch on the diagonal or form a diamond network. These forms will, of course, be especially valuable for trellis patterns but they also underlie many all-over patterns.

The principle of organization by the continuous growth of lines is the basis of most of Morris' own designs. He believed that it was by far the highest type of arrangement. He derived the idea from his Gothic study of art. In Gothic pattern he found that interrelation of curved lines developing from each other by the necessary force inherent in the lines bound together the parts, instead of the commoner and more obvious devices of mere contiguity or interlacement that are used, for instance, in a great deal of Renaissance and Classic design. When lines do spring from each other by their own power of growth, the resulting pattern will not take the more obvious geometrical shapes. This is especially desirable in wall-paper design, because the paper is going to be spread flat over a large unbroken space, and the skeleton must not show through too clearly. Moreover, you will, of course, on this so-called base of design, have a much freer development and a much wider range of possible patterning. Only, the danger is that freedom may become license and the lines spring with so little force of necessity that the pattern becomes over-complicated and confused. The designer who is going to free himself from the more rigid substructures must have a sensitive feeling for the quality to prevent himself from wandering at random and so getting lost.

The effect for which the wallpaper designer is striving is a pattern in silhouette. A design that plays entirely on mere color masses without specific form or outline, unless it be a blend which, in this sense, is outside the realm of pattern, is not appropriate, because it belies the formality of a repeated pattern and denies the mechanical technique of the production. Wallpaper printing is a kind of engraving and the basis of all engraving is line. To omit line and resort direct to color in a sort of limitation of painting is to lose the values of the one art without gaining those of the other. Any art that oversteps its own technical limits, no matter how remarkable the virtuosity, sacrifices its own qualities and is at best only an imitation of the other art. So wallpaper being one of the engraving arts must rely fundamentally on line. But, on the other hand, a wallpaper design cannot depend entirely on line, for that alone would not have structural strength enough, nor would it have sufficient mass and continuity to cover the wide surfaces. Mere line would also fail to carry across a room. So wallpaper design must, like its parent art tapestry, take the intermediate resource of silhouette, color areas clearly defined by outline.

Because it is an art of silhouette the quality of the outline in a wallpaper design will be all-important. The outline need not, in all cases, be actually drawn—color boundaries in themselves constitute lines—but whether drawn or not it should be clear, clean and definite. A blurred outline will lose the clarity of silhouette, making it an uncertain approximation to a color spot, and confusion of aim is always distressing in any art. A vague outline will be baffling and irritating. To avoid indefiniteness, it is desirable when there are a number of colors overlapping to have the boundary lines actually drawn to keep every color in control.

When the outlines of a design are drawn they may be either regular lines of an unvarying width and heaviness, or they may vary in width, increasing and decreasing to meet the needs of the design. The regular, even outline is essential in geometric and architectural patterns and sometimes in the more conventional renderings of natural forms. The varying outline is usual in naturalistic designs. But whichever type of outline is used it must be used consistently throughout the entire design. A formal pattern that includes some flowers, for example, cannot be drawn through the formal parts in regular even lines then have the flowers rendered in sweeping, irregular outlines. They too must conform to the regular outline to be consistent with the rest of the pattern.

An outline can often be emphasized without

complicating the pattern by using a white marginal line. That is, the blocks or cylinders are cut in such a way that the color falls short of the outline by a minute fraction of an inch, leaving an infinitesimal white strip between the color and the outline. When this is done, however, the space must be regular enough and wide enough to be clearly intentional so that it will not appear as a defect in printing.

Another type of emphasis that can be used is the white break at the junction of outlines, a particularly happy device because it makes clear the technique of the printing. Still another variation in the effect of the outline can be obtained by taking a hint from Gothic tapestries and instead of the usual black outline using a brown line which gives a softer and richer effect. Again, when the color of the design falls against a background of another color instead of against a white or neutral ground, the transition to the outline can often be made with better results through a narrow margin of a lighter tone of the same color. That is, if a red rose lies against green leaves, a little line of lighter red can with advantage be left just outside the outline of the rose.

In a design using human figures, such as

many of the landscape papers and some, also, of the more conventional and repetitive patterns, the requisite clean, strong silhouette can be obtained by presenting the figures in their broadest, flattest attitudes. In devising these attitudes Greek bas-reliefs and Gothic tapestries are the most suggestive models. Of course the rule would seem to require that every personage in the design stand either directly facing or directly back to the spectator. But this would make a monotonous and inflexible effect. The designers of the bas-reliefs and tapestries knew how even to distort figures to vary the attitudes and yet keep the breadth of outline, without having them appear unnatural. So a man in profile will have a body almost if not quite directly facing the spectator, yet so cleverly is the adjustment managed that the distortion does not seem uncomfortable. It is a nice bit of technique worth observing.

The lines of a wallpaper design must have, aside from their representative function, aside from the part they play as stems or tendrils or outline of leaves, vitality in themselves. They must never be vague or wandering, uncertain in curve or direction. They must never become exhausted in a faint trail or waver, nor must they keep on going when they seem to have consumed their original impetus. Even when a line ends it should look as if it had still great capacity for further growth. When one line builds on another in a continuous pattern the lines must never appear to sprawl, nor should the pattern grow too far from its original support so that the sequence is in danger of falling apart. All these injunctions are of the practical wisdom of William Morris.

In the drawing of curves or arabesques the wallpaper designer can get valuable hints from old Persian carpets, especially those of the Sixteenth Century. In, for instance, the cloud band patterns on Sixteenth Century Persian carpets, the curves have a vivid vitality. This comes from their continuously asymmetrical development. In perfunctory drawing we tend to fall into a regular symmetrical form. The same line drawn with interest and verve, on the other hand, shows new movement at every moment and so, in the case of a curve, develops a rapidly changing system of tangents. It is this energy of line that makes one of the great differences between vital design and mechanical design. The degenerate Ispahan carpets of the late Seventeenth Century illustrate the point by contrast; for in these, the same cloud bands that in the pieces of a century earlier had snapping movement now are dead and sluggish, following a regular, absolutely even curve.

Above all, in wallpaper design, it is important for the paper not to develop any accidental forms or lines, when the repeated pattern is placed on the wall, that the designer did not originally intend. This happens often in the ordinary wallpaper. Continuous stripes, vertical, horizontal, or, most common of all, diagonal, appear where none was meant to be. Or in some cases the lines of the pattern enter into new combinations and relations to form unexpected objects. A cat sitting on a back fence is revealed where the curve of a tendril touches the trunk of the next vine, or a hideous old witch peers out from behind a rose cluster. These secret worlds in the wallpaper can be very entertaining for the sick-a-bed but they can be maddening when they haunt the rooms day after day. Charlotte Perkins Gilman has a very convincing story of the insanity that came from an insinuating pattern on the wall.

Since wallpaper must, in general, be seen from a distance and serve as the background for other decoration it is desirable that the pattern fall together into rather large masses. Little bunches of flowers, for example, must be contrived so that they can merge into a general color effect when seen from across the room. If a stripe is decorated with a chain of field flowers, that must appear as a stripe of a general tone when the pattern is too far from the eye for the particular blossoms to be distinguishable.

But though the pattern must be effective in mass it should at the same time profit by rich and delicate detail. Since wallpaper design is a linear art it holds within it the possibilities of very nice detail. Minor incident in rich abundance will enrich the effect of the whole, provided it is not allowed to over-complicate and so confuse the main outlines. And it will be less likely to confuse the silhouettes if it is cleanly and crisply drawn as it should be.

An important device for enriching the design of a flat surface without destroying either its flatness or its mass effect, and also without losing its unity, is the pattern on two or three planes. Here again Persian carpets are the finest prototypes. The main design in the great carpets of the Sixteenth Century is often carried out on one scale and in one set of color

Victoria and Albert Museum

TEXTILE PATTERNS WERE THE EARLIEST USED IN WALLPAPERS. FACSIMILE OF PORTION OF EARLY WALLPAPER FOUND IN MASTER'S LODGE, CHRIST COLLEGE, CAMBRIDGE. SIXTEENTH CENTURY.

Metropolitan Museum

ENERGETIC CURVES ARE NOWHERE MORE VITALLY DRAWN THAN IN SIXTEENTH CENTURY PERSIAN RUGS.

values. This design makes in itself a completely interrelated whole. But it does not fill the entire surface. Secondary and supplementary to it is another pattern carrying out the same motives and the same general movement but, usually, carried out in a somewhat contrasted rhythm and in a smaller scale and lighter color values. This, too, is a continuous whole, but it also serves to weave closer together the main design. Sometimes, also, there is a third intermediate level of sparser pattern that plays through the mesh of the other two, echoing them. But the third level is extremely difficult for the designer to handle without producing confusion. The two-level pattern, however, creates a full and interesting decoration, helps to make a large scale design less obtrusive.

The scale of a wallpaper pattern is determined partly by architectural requirements, and partly by the mechanism of production. As far as the latter is concerned the unit of the pattern can be as small as the designer chooses. There are, for instance, stripe patterns printed in which the outline of the stripe is indicated by a series of dots no larger than periods. But the maximum size is strictly limited, the usual cylinder being from five to seven inches in diameter, permitting a repeat no more than twenty-one inches long, with occasional exceptional cylinders nine inches in diameter, permitting a repeat as long as twenty-eight inches for manufacturers who have oversized machines. This technical limitation accords with decorative requirements, for a larger scale pattern would dwarf any but an unusual room with furniture on the palace scale. The minimum sized patterns, however, that are possible with present methods of production really fall below decorative utility, being, in all but exceptional cases such as that cited in which the series of dots function as a line, too small to give any mass effect from across the room. The result is an appearance of chaos and confusion. Too small a pattern can, too, be just as over-insistent as too large a pattern because of the frequency of the repeat and the effort of attention required to interpret it. The possible scale of a useful repeating wallpaper design ranges from about three to twenty-four inches.

While a wallpaper design may employ the full range of colors, and indeed many of the fine landscape papers needed almost the full rainbow, it is unwise in any case to use more than three or four values of one color. A greater range of nuances brings the pattern out of the proper limitations of printing into competition with painting. Moreover, a complicated range of values in the pattern creates too many planes and this conflicts with the flatness of the wall. And, also, the multiplication of values confuses the design, preventing it from falling into the broad masses that are desirable.

William Morris is particularly emphatic in insisting that when a dark tone is needed for a paper it shall be obtained by graying the pigment, not muddying it. He also suggests that shadings be done by hatchings as in tapestry, that is by running one area of color into another with an intermediate zone of a comb-tooth pattern, as it were. This maintains the linear character of the art and, at the same time, insures every area a complete outline so that it stands out clean and sharp and does not seem to run into the next color.

In designing a wallpaper frieze the frieze may be regarded either as the dominating decoration of the room or merely as the border of the paper. If the frieze is regarded as the most important element of the decoration it can, of course, be larger in scale, richer in color and more complex in design. In either case, however, the main motives must echo the design of the accompanying paper or be echoed in it or, if the accompanying paper is plain or a blend, the colors must echo from one to the other. When both frieze and paper are patterned there must be the same quality and rate of movement in both designs. The movement in any frieze must have a clearly defined rhythm and a well carried continuity.

In general, the problems of wallpaper design are the problems that arise from the necessity of meeting opposed requirements. The design must be clean and definite, but it cannot be obtrusive. It must be crisp in detail, but must fall into large masses. It must be limited in the size of the unit, but must be continuous over an indefinite area. It must be pure in color, but must not be glaring. It must, in short, be at once a background and a positive contribution to the decoration of the room.

CHAPTER VI

THE IMPORTANCE OF WALL DECORATION

IN the last fifteen years or more decorators have tended to efface rather than to decorate the walls of their rooms. They have been inclined to make the mistake of thinking of the room as everything inside of the walls and the walls only as the inclosure that happens to be necessary. So instead of using them as an opportunity to give interest and character to the scheme, they have dismissed them as background and wiped them out of the picture by clothing them with an unbroken and uninteresting expanse of vague, unpatterned color. Thus the walls, obliterated, have dropped away from the room, leaving all the burden of decorative quality to the drapes, the furniture and the objects of art, and they alone have had to carry out the effect, unsupported by an actively functioning architectural setting.

Now as a matter of fact the walls of a room, far from being a mere unavoidable necessity of construction, are a positive feature of the room, of primary importance to its effect. They are the defining limit of the room and as such they give it its basic character. From one point of view the walls are the room. They make it and in making it fix its quality. If you simply blot them out, therefore, with the least interesting and so least conspicuous covering that can be devised your room will lose in strength and character.

As the defining feature of the room the walls should set the key for the whole decorative scheme. They should announce the basic type and character of the room in all the major aspects, be the opening bars, as it were, presenting the theme and forecasting its development. In color they should indicate the general range of values that the room will occupy. In texture they should record the feeling to be sought. In scale they should establish the unit of measurement. And these three, color, texture and scale, together with the pattern if there be one, should express the pervasive quality of the room, whether it be formal or informal, gay or dignified, personal or remote, conventional or capricious.

But even when the walls are not regarded as

the defining limit and so the essential factor of the room, but are used rather as a background only, they cannot therefore be summarily eliminated from the scheme with a negative treatment. For the background is not outside the picture. It is a very positive and very active element in the picture, a real and by no means minor factor in the final success of the whole. When the walls are considered primarily as the background of the set, as it were, their treatment will follow instead of lead that of the rest of the room, but it will still have to be a positive treatment, not the customary casual annihilation by means of blankness and characterless color and texture to which the usual decorator resorts.

The walls even as background will be unavoidably important for the simple reason, if for no other, that they are the largest in extent of any decorative feature. Twice or more as big as the floor, they cannot be entirely overlooked, no matter how vague and flat a treatment is devised to minimize them. And, too, they are omnipresent. Even in their least conspicuous raiment they insist on entering into every effect within the room, conditioning every other feature in it, making a decisive part of every group. The curtains, the rug, the furniture, all gain or lose by their relation to the walls, and every arrangement, even of the lesser pieces of furniture, has to take account of their influence. A grouping, for instance, against a dark wall should leave different spaces from that against a light wall, for a bright background makes a more insistent demand on the attention, so that a smaller space higher in key counts more heavily than a larger space of lower value. Again, an arrangement against panels must balance with the design of the panels, and this design will, in turn, hold together and emphasize the group, giving it a sharper and more highly individualized effect. Also, a composition of pieces of furniture in a room with a patterned wall must be made with attention to the lines and main points of interest of that pattern in order not to run counter to them, and when skilfully done it will utilize the design as a contributing part of the group. In short, whatever the problem of decoration, whether small or large, the walls and their treatment are an active factor in the statement and solution of it.

In all the great periods of decoration the walls have been a positive factor, carefully and often elaborately treated. In the Gothic houses of England the panels, thoughtfully proportioned, with beautiful moldings and often beautifully carved, were brilliantly painted in gold and vermilion and other strong colors. In the Tudor houses hand-troweled plaster often took the place of panels, but in many other cases a hint was taken from Italy and Spain and rich damasks or finely wrought leathers were used. The late Seventeenth and Eighteenth Centuries was the culminating period of beautiful walls both in England and on the continent. Then wallpaper came in with brilliant and varied patterns painted by the Chinese. Then silk wall coverings, too, came to an even more luxurious height. And panels were painted by great masters or heavily carved, or, at the very least, delicately adjusted with thoughtfully planned moldings. Think of the work of Grinling Gibbons, of the Brothers Adam, of Boucher and his followers. Wall decorations called upon all the decorative arts, even the arts of the metaller in some of the elaborately wrought frames for panels and overmantels. It was really an art and a master of the arts.

The great houses of America have for the

most part maintained the continuity of the great tradition. They have had wall decorations of strength, character and individuality, because they have been made by architects who approached the decorative problem with the architectural interest that emphasizes the significant rôle of the walls. But in the less costly establishments done by the average decorators, even though they be quite ambitious, and from these down through the simpler levels of suburban homes, the cult of the mild still prevails. So we have house after house with plaster walls stripped bare and blotted out with a flat neutral tone that fails to function as a color at all, apartment after apartment with paneled walls faintly indicated with undecorated moldings and smeared with some dull and characterless paint, and room after room covered with colorless plain papers of one popular kind or another. And even what pallid colors the decorators do use are not allowed to do their best but must usually be "antiqued," commingled with some dust or smudge to rob them of any freshness they might have had, and finished with a queer, gluish, dead gloss. So that what should be Adam green—the pale, sharp, bluish pea-green —is a sticky, grayish tan; and what should be fresh and white looks as if it had been left to soak in a vat of weak and unstrained cold coffee. Sometimes it seems as if most of the apartments and houses of America have all been lost in a London fog, a dirty yellow fog, and all, also, in the same fog. Everywhere you see the same dull and undistinguished walls. The milliner's and the dressmaker's shop, the tea-room and the theater, your friend's house and your own are all clothed alike.

Current wall decoration first lost touch with its fine tradition in a reaction from the horrors of the middle-Victorian age. Heavy, muffling drapes with gross patterns, contorted scrolls and leaves in violent chemical dyes on wallpapers and fabrics, the monstrous convolutions of jig-saw moldings; all these provoked such a strong antipathy in the younger and more enlightened generation that they made the mistake of fleeing to the other extreme to correct it. It is a usual human error to assume that because a thing is bad the good would be its complete opposite. But this error overlooks the fact that the worst kind of ugliness is perverted beauty, the most distressing ornament that which fails but which might have been good if done right. So the corrective of bad decoration is often not the most remote alternative conceivable but that same decoration properly thought out and carried through.

The reaction away from decorated walls was partly due, too, to the fact that this younger, suddenly illuminated generation, awakening from the bad taste of its parents as from a nightmare, felt a little uncertain of itself. If such horrors could have been endured by their parents with no awareness of how bad they were, with pride and enjoyment, even, how could they feel quite certain that they were not being equally blind to their own deficiencies? Thus intimidated they had recourse to safety as their guide. And pattern was avoided first as the most disastrous pitfall. It was pattern that had been the worst offender in Victorian decorations. Beware pattern, therefore. Extol the beauty of plain surfaces. Raise the standard of simplicity.

But even plain surfaces can be bad if wrongly colored. Those who had been brought up with the Victorian reds and greens knew that. Better, then, play safe with color too. Patronize the unoffending neutrals. Depend on the surely unobtrusive soft tones. With these you can make no bad mistake and if you make of them a cult you can even become accounted a person of sophisticated taste.

Such was the unconscious reasoning at the end of the last century and the beginning of this when our present vogue of the innocuous was established. It was the product of the lack of self-confidence, of fear of the undeniable difficulties in the successful use of pattern and real color. And once the trend was marked out the attitude found reenforcement from many sources. Among these was the sentimental æstheticism of the '90's. This movement of over-nice young men and women fondling a talent for one of the arts expressed itself in soft tones and grayed colors, general vagueness and weakness. Crudity was the greatest sin, and strength is easily named crudity by those who do not possess it. So the strength of pure color was taboo.

Further support was found in the art of Whistler and his imitators. The conspicuous eccentricity of the man, the literary interest in him, helped establish and helps maintain the prestige of his Symphonies and Nocturnes. Here is an honored prototype of grayed tones and blurred effects. But the Whistlerians, conscious and unconscious, overlook their master's aim in using color of this quality, and forget the different resources he employed when he undertook to decorate. The pigments and technique necessary to render certain light and atmosphere effects has little real bearing on the quality of color in decoration, and when Whistler rendered a mural design in his famous Peacock room he used, not faint and fading paints, but pure strong colors with decisive pattern, and used them in abundance.

The vogue of antiques, too, has gone to maintain the fashion of dull colors. Here, too, of course, their exponents have builded on a mistake. In part they have not realized that when an antique fabric is dull in tone it is because the ravages of time have robbed it of some of its proper beauty. But the growth of expert knowledge of the arts in America is demolishing this fallacy fast. The frank exuberance of the Gothic has been shown to have been extended to their use of color, too, and finely preserved examples of their polychrome appear from time to time to emphasize this. The cool interiors with their plain walls that so delight us in the work of Vermeer and his contemporaries, were richly decorated with maps, relieved by intricately mullioned windows, strengthened by bold patterned tiles and enlivened by the deep glowing colors of the ever-present Asia Minor rugs. If the Eighteenth Century was delicate it nevertheless was gay, and even the pure white temples of the Greeks are now known to have been unashamedly garish. There is little comfort in the past for the color-timid souls. But in part the exponents of "soft" colors, even when faced with these facts, have evaded them by sentimentalizing over the very deterioration incident of the passage of time and wear, and so have treasured things in proportion as they were dim with dirt and heavy use.

The biographer of William Morris has preserved a characteristic and valuable anecdote. A titled patron of importance, having seen some of Morris' work, came to him to have him do over his drawing-room. Now the fabrics that the client had seen and admired were some of those which the firm had made before it undertook to do its own dyeing, when it was trying to use commercially dyed wool. This Morris had found unsatisfactory, for he had never been able to get the pure, clear colors on which he insisted. So, when his titled client came, he had recently constructed his own dye house and was now getting the strength and cleanness of color that he desired. The customer was shown sample after sample of these properly dyed carpets and rejected each in turn. The colors were in every one too strong. At last he had seen all the colors and types that were being manufactured by the firm and still he was unsatisfied. "Is this all?" he asked. "And where are the nice soft tones that you used to make that I saw in Lady Blank's drawing-room?" Morris had gradually become more and more enraged, "Sir," he thundered, as he started to walk away, "if it is dirt you want go out in the street and get it!"

Unfortunately the taste for dirt that is so prevalent now reenforces itself in a continuous vicious circle. For the appreciation of clean and genuine color and of exactly drawn pattern can develop only when there are such color and such pattern to be seen, and, on the other hand, taste can rapidly degenerate and weaken through the habit of seeing muddy colors and blurred designs. Only a sharp, spontaneous recurrence of vitality can break through the endless chain and reestablish decoration in a fresh, direct medium. There are some signs that such a regeneration is about to appear. Color has begun to come back into its own and pattern is slowly coming into favor again. But a habit of taste is not quickly or easily broken. When a combination of timidity and false theories, and the anxiety for correctness which the two bred, muddied our colors and enfeebled our patterns, an evil tradition was inaugurated that will be difficult to eliminate.

Until we do succeed in shattering the decorator's fetish of softness and unvaried neutrality, which really means vagueness or emptiness, we will have a type of decoration robbed of all of the essential qualities of decisiveness, architectural exactness, and individuality. That is to say, we will have a type of decoration that does not solve any of the major problems but abandons them. When, however, we do get again the courage to wade without hesitation into the basic decorative elements of color and design, and struggle with them at their strongest, we will again have an art of decoration that can take its proper place in the historical sequence.

What is more important, we will have again homes of distinction and individuality. At present, we are all clothed in the same uniform, or so much alike that there is little to tell us apart. And this is necessarily the case, for neutral colors without design give no range from which to choose, and even what difference there is between tan plaster and a gray paneled wall are too slight to create a decisively different impression. But in color used with courage and in design there is a medium of expression that can be nicely attuned to every different need, taking our houses out of their present repetitious monotony. With the return to walls that are positively decorated instead of just effaced, we can begin to get once more a robust expression of individual character and taste.

CHAPTER VII

COLOR PROBLEMS

MORE nonsense has been written and talked about color than about almost any other phase of familiar experience. For some reason it has caught the fancy of many types of intellectual charlatans. Turned pseudo-psychologist they have victimized the interested public variously. Seizing on small facts they have made big statements, written whole books even. Their fantasies have ascribed all kinds of sinister possibilities to colors and their combinations. They have dilated on their emotional qualities, their symbolism, their therapeutics, their relations, the music of color, and so on, and they have perpetuated so many evident and silly misstatements that now any discussion of the subject is at once suspect.

But in spite of the vagaries of the color mystics, there are certainly definitely established facts about the psychological effect of colors and of color combinations that are of interest and value to the designer of interiors, whether professional or amateur. Color is so important an element of decoration that any verified facts about it are interesting. And while in the use of color the real artist can rely on his unanalyzed sensibilities, and in the end no facts can ever entirely supersede the need of good taste, definite knowledge can always guide and clarify the vaguer feelings. The specific evidence of psychological experiment on the effects of colors and of color relations can reenforce instinctive preferences and codify traditional usages.

Colors do in themselves have a quite definite effect on the average normal person. The artist discovered this before the psychologist and recorded his observations in his long-established distinction between the warm colors and the cold colors. The warm colors, as every painter knows, are those at the red end of the spectrum; the cold, the green and blue end. Now this old distinction is no mere arbitrary characterization. It does correspond to an actual experience. The usual person will, when surrounded by pure red color, feel stimulated, warmed, excited even, and the feeling comes from an actual stimulation of fundamental physiological processes—increased heartbeat, increased respiration, heightened muscular tension. Similarly the blue-green end of the scale does actually make the usual person feel chilled and depressed, for it does express these same physiological functions, does tend to retard heartbeat and respiration and decrease muscular tension. The middle of the spectrum, the yellows and green-yellows, are emotionally indifferent, though pure yellow by the association with sunlight usually seems warm. The so-called neutral colors, the tans and grays, take on one characteristic or the other, stimulating warmth or depressing coldness, according to the dominant tone in them, for tan and gray are seldom pure colors but are usually suffused with one of the spectrum colors. As tan more often has a yellow or red component it is usually warmer though there are cold tans bordering on gray. Gray may be very cold as in the blue and violet grays, or quite warm as in the pink and orange grays.

All this is, of course, significant for the decorator. A cheerless room can be given a different quality with warm colors, or, if neutrals are needed, with warm neutrals. A room too brilliant can be tempered with cold colors. A room that must be as quiet and undisturbing as possible can safely stay in the yellow range or yellow grays.

Probably because they are exciting and interesting the warm colors are more intrusive; the cold tend to recede. So a room with reddish yellow walls will look smaller than, say, the same room with light green. Similarly the more intense tones of a color, the more highly saturated in the exact terms, approach the eye; the less saturated withdraw. And, in the third place, the lighter values of colors, the tints that contain more white, also tend to recede. Therefore the small room looks largest with a pale, light blue or green paper and smallest with bright, dark red walls. This recession of the high-keyed and lightly saturated colors is probably due to the fact that as an object gets farther from us the color of it appears lighter and paler, until at the farthest point of vision it looks like a whitish gray blur. So we are inclined to interpret the object that is light and faintly colored as farther away, that which is dark and more intensely toned near. This is the principle of aerial perspective so effectively employed in Chinese and Japanese painting.

So we unconsciously adjudge the walls of a room to be farther away when they are pale and high in key and thus the room seems larger.

Another aspect of color to be taken into consideration in the choice of a wallpaper or any other decoration is the effect of the color on the light and of the light on the color. A paper pale and light in key, of high value, will reflect more light and so make a room lighter than a paper dark in value, because the dark color absorbs some of the light. That is, in fact, what is meant by value. The value of a color is high in proportion to the amount of light it reflects. So a dark room will become brighter with a light paper. But some colors lose their luminosity more rapidly in diminished light than others. This fact, too, must be taken account of in the selection of a color for a very dark room. In a dull gray light red rapidly loses its brilliance and even its color whereas green remains bright. In a hall with no direct light the clearest, most colorful paper would be a greenish yellow.

A pure and quite fully saturated color is to any discerning person more pleasant than a faint or muddied color. The commercial dyes that were prevalent until recently dulled our appreciation of color and perverted our taste, and the perversion has been exaggerated by the prevailing popular theory that "soft" colors are more "refined" and in better taste. But if you can get an honest unbiased expression of preference most people will turn to pure color quite fully saturated. We enjoy seeing anything in the most complete and perfect expression of its own characteristics. So a finely bred prize bull that is everything that a bull most typically should be gives us æsthetic pleasure. The full realization of the type in any individual has strong intrinsic attraction. It is this that leads us to prefer the pure and saturated tone, for in such a color color is most fully and truly itself.

Saturation and purity are very important in red, blue, and yellow, and the natural preference for these at full intensity has expressed itself in all the great periods of decoration that did not suffer from excessive civilization: the Egyptian, the Greek, the Gothic. Full intensity is especially desirable in the emotionally indifferent colors at the middle of the spectrum —the various shades of yellow—for since they do not in any case affect us strongly they must be seen in their utmost intensity to arouse any great interest. Green and violet, however, are rather more pleasant at the medium saturation point.

But if this preference for full saturation and purity were given full rein in our decoration our rooms would be too violent and too exciting. As tired a world as ours cannot support as interesting an environment as the more robust epochs have chosen. The problem, then, is to satisfy the need for pure, vital color and still quiet the effect. The solution is the grayed color which must be kept distinct from the thickened or weakened color. The grayed tone retains purity with its softness and seems to us to be still strong and vital because, according to one noted psychologist's explanation * we unconsciously consider the gray cast a veil between us and the real color which seems stronger than ever since it has the power to glow through the veil. This impression is enhanced if there are in the decorative scheme a few touches of the pure ungrayed color so that it will seem at spots to break through the veil. Thus in a room with a grayed blue paper some smaller decorative objects of the pure blue will give greater vitality to the paper without making it seem any less reposeful.

* Theodore Lipps, Asthetik.

There is, however, one danger in grayed papers for the walls. The veil appears rather as an atmospheric veil, and if the paper gets too grayed, too atmospheric, the walls are robbed of their necessary solidity and seem to float instead of to support the ceiling firmly.

When it comes to the combination of colors the facts are more complex. In the first place, the number of distinguishable colors is enormous. We simplify the world of color by reducing it to six main divisions that we consider—red, orange, yellow, green, blue, and violet or purple—and in a sense these are fundamental. But there are actually one hundred and fifty distinguishable colors in the spectrum series. Besides these spectrum colors every one of the hundred and fifty has a large number of different tints created by varying admixtures of white or black. Blue, for instance, or any one of the blue-greens can range all the way from a tint a few degrees above black to a light tint almost as high in value as white. The multiplication of these tints increases the number of distinguishable colors to between forty and fifty thousand. In addition to these variations there are numbers of other colors composed of combinations of colors not appearing in the spectrum such as the purples, combining blue and red, unusual greens with an admixture of red and so on, these being generally called in the exact terminology of psychology shades of the colors. Each of these shades also has a full range of tints. It has been estimated that with all possible combinations in all possible values there are between five and six hundred thousand different colors. The possible combinations among these are of course almost incomputable.

Nevertheless certain general facts about color combination can be asserted. Every color has its own complementary. This complementary is the color that is its exact opposite, and the sum of the two when mingled by any one of several mechanical devices is white. This does not, of course, mean that pigments of these colors when mixed would give white paint but that when the two colors are seen together the eyes see white or pale gray. A simple test experiment to find the complementary of a color is to stare hard at an intense spot of that color then look quickly at a plain white surface and you will see a spot of the complementary. The eyes tired by seeing one color too long and hard find compensation in its opposite for some reason not yet fully determined. The complementaries of some of the more important colors are:

Red—green-blue
Orange—blue
Gold-yellow—blue
Yellow—indigo
Green-yellow—violet
Pure green—purple

The most immediately pleasing combination of colors has been found in repeated experiments to be the combination, not of the exact complementaries, but of shades approaching the complementaries. The complementaries closer to the dividing line between the warm and cold colors, a line that runs between yellow and yellow-green, are more pleasant in combination than the complementaries more remote from this, as for example the green-blue and red combination.

This does not mean, of course, that in decoration it is always most attractive to combine complementaries, for the effect might in some rooms be too interesting and too strongly pleasing for the quiet feeling sought. But it does mean that for a strikingly pleasant effect the complementaries, or rather near complementaries, are a sure device. So yellow wallpaper with dark blue curtains has a vivid attraction; and with a greenish-yellow paper violet would be entirely delightful, a less often exploited combination.

In combinations between colors not in the complementary relation, the indifferent middle range of the spectrum—the yellow-greens—combine successfully with the widest range of other colors. Red and purple are the most difficult to use in combination since they combine pleasantly with relatively few other colors.

The preference for colors against a dark background is quite different from the order of preference of colors against a light background. Against a dark ground red is the first preference of the usual person, with yellow, green, and blue in descending preference, whereas against a light ground blue is most attractive, with red, green, and yellow in descending preference. So with a dark wallpaper minor decorations emphasizing tones of red and yellow would ordinarily be most pleasant whereas with a light ground the relieving bits of pottery and other ornaments should, to get the best effect for these minor objects, be less yellow, more blue and red.

In combining color with a neutral gray or tan the indifferent colors of yellow and green-yellow are the least interesting, the more emotionally positive colors, red and blue, more interesting. There are, of course, some occasions when a negative room is desirable, a room not impressive or moving in any way. For this a yellow and gray combination would be perfect. The tints of any color—that is, the tones that are lighter than the pure spectrum color—combine better with the neutrals than the pure spectrum color. But the pure, true, spectrum colors combine more satisfactorily with the neutrals than do the shades—the tones resulting from an admixture of black with the spectrum colors. That is, a pure blue in a rather high key is pleasanter with gray or tan than a dull shade of blue or purple. But any color with gray is pleasanter when in quite full saturation.

For less vivid combinations, colors nearer together in the scale than the complementaries are more useful, but the colors must not be so near together that they are not readily distinguishable. Two shades that almost but do not quite match are thoroughly unpleasant together, partly because they demand from us too much effort to tell them apart, partly because they are apt to seem like an unsuccessful match. An actual case of this mal-admixture, more distressing than any that could be readily imagined recently appeared in broad daylight: a turkey-red hat trimmed with a broad cerise ribbon that fell down across a purple-red coat. It was an appalling apparition. This is the worst possible example but there are less striking cases of the same kind of disharmony that occur in two-toned wallpapers and fabrics and which it is important to avoid.

Among the most successful of the less striking combinations are those of a color with a tint or a shade of itself, the pure orange of the spectrum, for instance, with a lighter tint of itself or with some of the terra cottas that arise from a combination of orange with some of the browns. The colors at the middle of the spectrum combine better with tints of themselves than the colors at either end. So we have the justly popular decorative scheme of browns with golden-yellows. Red, however, with a tint of itself, or violet with a tint of itself is quite impossible. Hence the traditional objection to red and pink. And violet and pale lavender are quite as bad together. Similarly, colors at the middle of the spectrum combine better with shades of themselves, such as orange and terra cotta, than with the colors at either extreme end, and the warmer colors in this range combine better with their shades than the colder, the oranges better than the greenish-yellows. But, again, some greens and purples are good with their shades. Blues and violets are on the whole quite unsuccessful in this kind of combination.

In using together a bright color and a dark color it is always more pleasant to have the bright color concentrated at central spots against the dark color which appears as a more extended background. This preference is reflected in the common decorative practice of using low-toned walls of less intensity against which small brilliant decorative objects introduce spots of color. Always we find it pleasantest to center the most brilliant. This means, too, that in composing a group of furniture and decorative objects, it should be balanced about the most brilliantly colored object, using this as a fulcrum, as it were.

For if some less brilliant spot of color is chosen as the balancing point, and the more brilliant thing pushed to one side, the eye will refuse to accept the disposition, and, fixing on the more brilliant factor as the balancing point, the whole group will be thrown out of equilibrium. One readily feels that the brightest, most intense color is the concentration point.

When all the colors in the group are equally bright the color nearest the red end of the spectrum will be chosen as the center. This will be somewhat modified, of course, if there is some object that, because of size, shape or pattern, is strikingly so much more interesting that it seizes the attention without regard to color, but other things being about equal the brilliance or the warmth of the color will determine the focal point.

Most successful color combinations are instances of the fundamental æsthetic principle of unity in difference. The organization of contrast and variety into unified effect is the basis of composition in all the arts. It is quite evidently the basis of line design, which uses all the devices of balance, symmetry, compensation and repetition to unify a variety constituted of different kinds, qualities and directions of lines, but it is equally essential if less evident in color composition. The basic preference for the combination of near complementaries is not so easy to understand as an example of this because the psychological relation of complementaries and the chemistry of vision are not yet clearly explained, but the white or almost white light that is the result of the combination of the two colors and that is the usual medium of vision seems to act as the semi-unconsciously present unity which holds together the two sharply contrasted colors. That colors slightly off the complementaries are more attractive than the true complementaries may be due to the fact that these arouse a little effort at correction, an effort that makes us realize the colors more keenly, just as in ragtime the slight failure to maintain the exact rhythm forces us to a correction that intensifies our awareness of the rhythm. The combinations of colors with tints or shades of themselves are simple cases of difference in unity.

The fact that all colors are not pleasant with tints or shades of themselves shows that all unity in difference is not necessarily pleasant. Nevertheless it is usually pleasing and there are several less obvious applications of the principle that should be more generally recognized by designers. The tendency, especially in interior decoration, is to rely on a few obvious types of very simple color schemes. Certain complementaries of which perhaps the most usual is gold and blue, and a neutral with one strong warm color, such as shades of gray with touches of orange, are endlessly repeated. Because they are so simple and also because they are so often used with only slight variations, these combinations readily become tiresome. Their obvious simplicity tends, too, to make the rooms look artificial and "decoratorish." The designer can profit, therefore, by considering some of the more complex forms of unity in color difference.

Two or more shades of different colors that have some color in common will usually combine pleasantly, the common color acting as the underlying unit to tie together the contrast. So a green-blue and a green-yellow, or a yellow-red and a yellow-green, or a blue-green and a blue-red are all possible color combinations. A particular type of this association of colors with a common factor that is very useful in decorating is the harmony obtained in a group of contrasted colors when they all have a gray tone. The gray holds them together, and underneath this veil, as it were, there can be an energetic play of strong and varied shades.

A less obvious connecting chain among a variety of colors is a common value. The value, or degree of light absorption or reflection of the colors, can well act as the common factor. So shades of blue and green or of green and red which in different values are quite impossible can be combined when at the same value. The room, therefore, that maintains one key can use harmoniously a greater range of colors and shades than one that has a wide range of values. So Louis XV and Louis XVI rooms, that are entirely decorated in a very high key, can, without seeming disorganized in color, use a number of different tones that is amazing when one begins to count them.

But association by a common factor is not the only kind of unity in difference possible in a color scheme. A more subtle kind of unity can be obtained, not by any common element, whether color or value, but by the order imposed by a gradual series. That is to say, two or three colors that clash violently can be held together by using as a connecting link another color that independently harmonizes with each of them. So a red and a blue that are not pleasant together can live together if associated with a yellow that combines with each of them. The series can of course be much more complicated with more members and more gradually spaced intermediates.

Similarly a varied color scheme in a varied range of values can be held together if there is a continuous sequence of values between the highest and the lowest, if there is a full range of the intermediate values, making a gradual but rather definitely stepped transition. In using this interrelation, however, the designer must have a care that his sequence does not fall apart again into sharp subdivisions, because all of one color and one range of values coincide and all of another color and another range of values coincide. That is to say, in a room with yellows, greens, reds, and blues, the yellows cannot all be keyed high and all the high values be yellows, with the next range of values identical with the blues, the next with the greens and all the darkest values confined to reds. If this is done the room is broken up into sections. There must be yellows at intervals all through the scale of values, blues at top and bottom as well as in the middle, several scattered values of the green and a few higher notes of the red. The colors, that is, must alternate in an irregular succession down the scale of the values.

The most fundamental consideration that should never be forgotten is that no combination of colors simply adds color number one to color number two to make effect one plus two, but in the combination each of the colors suffers some change. Each color is, as it were, to some extent mixed with the other color alongside of it. This mixing is done by the eye, which carries over some of the effect of the first color looked at to the second color. The eye sees the second color through a film, as it were, of the complementary of the first color, so that it sees that second color, not as it is in itself, but as a fusion of itself and a faint tint of the complementary of the first. This fact can be taken advantage of to increase the beauty of both colors, but when not taken into consideration the result may be a loss of quality of both colors that when alone were altogether delightful. So orange next to red becomes more yellowish and loses in intensity and purity, whereas it gains in brilliance when next to green-blue, because the complementary of green-blue is red, hence the orange becomes redder and so more brilliant. When any color of low intensity is put next to its complementary it becomes in this same way apparently more intense. So a rather weak golden-yellow porcelain will look clear and brilliant in front of a dark blue textile.

When two objects of the same color but of unequal intensity are put side by side the lighter color will look gray. So a fine apple-green porcelain against a dark green wall would lose seriously in clarity and intensity and look weak and grayish. To get the strongest effect from it it should be seen against a lavender-gray wall. When two different colors that are not complementaries are put side by side and they are of different intensities the paler color will suffer the most. So a chair covered with a fine pale blue French brocade on a dark, strong green carpet will look flat and ineffectual. Also, a smaller area of color is more affected than a larger area.

By a similar transference of the complementary, when colors are seen against a gray ground they enhance that gray ground by carrying into it their complementaries. So objects of art of bright colors in an all-gray room not only enliven the room by contributing their own spots of brightness, but, if they are large enough, enliven the gray itself. They will be more effectual in enlivening the gray ground, however, if when the minor objects are red or yellow they are of a lighter tint, a higher value than the gray ground, whereas when they are green or blue they should be a darker value than the gray. When there is no question of increasing light reflection, therefore, a darker gray wallpaper is best for a room with red and orange ornaments, a lighter gray for one with blue or green draperies and upholstery and decorations. Then the gray will absorb the utmost vitality from the other colors.

The wallpaper of a room can play either one of two parts in the color scheme. It either can be a decisive, decorative feature, or it can be merely a background. Whichever it is it should in color value be somewhere near the middle of the scale of values in the room. The floor and most of the furniture will be lower in value, the ceiling and the sharp decorative accents for the most part higher. But if the wallpaper is an actively contributing factor in the decoration it can be much more positive in tone than if it is only a background. As an important part of the decorations it should proclaim the dominant colors of the room and determine all the other color selections. As background it will, of course, be more neutral, taking the lead only in its quality of warmth or coldness. But whether positive or neutral it should always be of a clear, clean tone with some vibrancy, some interest, if only of texture, some reason for existence in itself.

CHAPTER VIII

THE IMPORTANCE OF TEXTURE IN WALL DECORATION

AN important element in the decorative effect of a room that is often disregarded is the texture of the stuffs employed. The qualities of the various surface finishes in a room will have a decided influence on the success of its design, an influence that will be felt in several different ways. This has been, however, quite often overlooked in recent years even by thoughtful and skilful decorators, so that many rooms that are well arranged and beautiful in color are quite haphazard in the choice and composition of textures, an oversight that detracts from all the other values.

The texture of the wall covering is especially important because in texture as in color and design the finish of the wall at once announces and determines the quality of the rest of the plan. A harsh strong texture limits in advance the selection of the wood finish, the carpets and the draperies. A soft, luxurious texture, on the other hand, will make possible only certain qualities of color and even a certain scale of the furniture and accessories. Yet even in spite of this decisive importance, the choice of a wall fabric is often made almost at random as far as texture is concerned.

The neglect of this factor of design is probably due largely to the fact that we do not realize how sensitive we are to texture. It does not intrude itself on our consideration partly because we have no nomenclature for it, because we cannot readily identify it with familiar terms as we can the colors, but must resort to adjectives to describe it and even they are limited in number and unspecific—smooth, rough, soft, and so on; and partly, too, it is not an insistent fact because so much of its effect is made, not independently, is registered not as texture as such, but as a modifying factor in some other factor, in some other aspect, an influence in changing the value of a color or even in modifying the apparent size of an object. But in spite of the fact that it is less obvious than the other aspects of decoration, texture is significant and we are very responsive to it.

How responsive we are to texture even though we may not be acutely aware of it can readily be seen by thinking of different materials with decidedly different textures. Think of marble, then of sandpaper, then of velvet. In surfaces of such characteristic quality as these we understand and appreciate the material almost entirely in terms of its texture. They mean to us the cold hardness, or the gritty roughness, or the yielding yet dragging softness, and if we are keenly introspective we can discover at the very mention of them a specific sensations in our finger-tips that they provoke. And even more do they provoke this sensation when we look at them. It is the same, too, though it is less easily detected, with materials of a less decided surface finish. We understand them partly in terms of the feeling that they give us when we touch them so that we enjoy them or find them disagreeable, largely, in proportion as we like that particular sensation. Think of the different effect on a cold day when one is very tired of a marble bench and a deep, overstuffed velvet couch. Yet the marble bench may in itself be a work of art while the couch may have little merit of design.

The European painter, especially the still-life painter, has long been interested in textures and has spent great ingenuity in the adequate rendering of them. In fact, the fine still-life painting is as much or more a composition in contrasted and compensated textures as it is in colors or space forms. And when we enjoy such a painting we are enjoying those sensitively suggested textures, even though we may not be definitely aware of that aspect of the picture.

So textures are influential even though we are not specifically conscious of them. We appreciate and react to them as textures, a fact that surely must be taken account of in employing materials in room decoration. But in addition to that, texture affects other factors in the decoration, too. Its influence on color is marked. The same tone of blue, for instance, in a dull finish, a napped material such as velvet, and a high glaze such as a Chinese porcelain, will not even appear to match, much less function as the same value. And, conversely, a quality of blue and green which if both were in high glaze would be most inharmonious may in a mat finish such as an old damask, or in a combination of two finishes such as the flat damask for one and a high porcelain glaze for the other, constitutes a delightful harmony. So variations in the kind of material require corresponding adjustments in the quality and values of the colors.

Similarly, texture affects the apparent size of objects. This it does largely by making them more or less conspicuous. A shiny jar that catches the light seems larger than an old leather box that has no high lights at all. Or, on the other hand, a velour couch with a deep nap seems larger than the same couch covered with a damask that is neither very shiny nor very absorbent of light. The strongly marked texture of the velour holds the attention and increases the apparent importance of the piece of furniture.

The effect of texture on color and on the size and conspicuousness of the surface is, of course, important in relation to the choice of a wall-paper. The former needs especially to be taken into consideration. The tone of the wall will vary markedly with the finish of the paper and this variation will be affected again by the amount, color and direction of the light that falls on it. In general, a highly glazed finish such as that used in a satin finish paper or in lacquer papers, or such as is created by varnish being applied after the paper is hung, will lighten the value of the color, and give strong illumination where there will be sharp reflections. Conversely, colors will seem darkened on a dull finish, especially if it is also a smooth surface. A broken surface like that of a crêpe or engraved paper, or even a slightly irregular surface like that of the grass cloths, will be somewhat lightened in value because there will be little bright spots of light thrown into sharp emphasis by little dark shadow spots.

The tone of the color, moreover, as well as the value, will sometimes be changed by a difference of texture in different lights. A blue-green, for example, on a ribbed surface may look decidedly bluish when the light falls on it from one angle, greenish from another. Certain paper designs take advantage of this effect and are planned to be a flat gray, for instance, from one angle, but decidedly rosy from another. This variation of tone that results from a play of light over an irregular texture gives the wall a little vitality and interest that is lacking in a flat, unpatterned color.

The influence of texture on size is of less importance in wallpaper selection than in furniture covering, though even here it cannot be entirely overlooked. A highly glazed surface is apt to make the ceiling look higher; a dull, broken surface like embossed leather to make it lower. Also, a design on a paper with a very decided texture such as a flock paper that has actual nap on it seems larger than the same design on a paper with little or no texture interest, such as plain, unembossed paper.

But in addition to the influence on color value, tone and scale, texture considered just as texture should be taken into account in the selection of wallpaper. For the texture should be chosen to be interesting in itself. A delightful textural quality is especially important for the wall covering of a room in which a dull unpatterned neutral color is necessary because of other decorative features strong in pattern and in color. The texture acts as a kind of pattern and serves to break up the flat, neutral tone, so that the wall gains some distinction and functions positively in the total effect as a decorative feature instead of being only an obscure background. But even a wall that is patterned can usually profit by having a definite tactile value.

For texture, skilfully used, has an expressive value of its own. A rough, decided texture increases the appearance of strength in a room. The structural significance of the wall is increased, its active work in supporting the ceiling. A rough paper seems to be part of the actual building material and so it makes the walls seem more compact, firmer, thicker, heavier. This was one of the values of the antique, hand-wrought plaster. It did increase the effect of solidity in a room, an effect especially important in the period when it was used, for the current furniture was heavy and massive and most of the accessories, andirons, candelabra, even table service, were large in proportion. With such furniture the strength of the walls had to be accentuated with a rough finish.

Similarly, in a room in which there are many books a rough textured paper is useful, for the books not only are, but look heavy so that a firm wall is needed to uphold them. Especially if the books are on open shelves is the decided texture valuable, for the books themselves then have a very marked texture effect, an effect that comes both from the varied materials of their bindings and from the heavy corrugation of the row as they stand pressed close together. A case of books acts, moreover, rather as part of the wall so that their marked texture must be carried on through consistently on the rest of the wall.

It is a failure to understand texture that has led to the passing popularity of the bare plaster walls. These have been used, apparently, in an attempt to recapture the character of the fine old plaster of the early English houses and of the Italian and Spanish palaces of the Sixteenth and Seventeenth Centuries, which did provide an exceedingly dignified and at the same time charming wall finish. But to strip off the paper from the ordinary plaster walls, treat them to a coat of coffee-colored paint, sometimes overlaid with a glue-like material to give them a dull, broken sheen, is not to begin, even, to regain the feeling of the antique plaster. For the essential quality of the real antique plaster comes from the fact that it was handmade and so had all the slight modulations and irregularities of the trowel to give it interest and quality. The flat, dead, mechanical surface of our modern plaster has no æsthetic value whatsoever, so that this completely destroys the walls as a decorative feature or, what is worse, makes them positively unpleasant by giving them a corpse-like effect of inertness which contradicts their structural function. Nor does the thin coat of grit that is sometimes sprayed onto the surface come any nearer to the interest of the old, hand-wrought plaster. In truth, where the real hand-troweled plaster cannot be obtained, if it is the essential quality of this plaster that is sought, the varying modulations and consequent broken lights and shadows, it can be better obtained in several types of strong-textured papers than in any mechanically finished wall.

As roughness of texture gives the feeling of strength, so the soft depth of a velvet surface gives the feeling of luxury and richness, a dignified, rather portentous luxury. In part, this may come from association, from the knowledge that velvets are expensive and used in dignified and rich surroundings. But, on the other hand, the very choice of velvets for occasions of state and solemnity might seem to indicate that they do from their own character convey some feeling of dignity and impressiveness.

Equally rich but less severe are the textures similar to satin. Taffeta and the silks of that kind, such as moiré, give a still more intimate luxury. The silk finishes with stronger textures, such as damasks and poplins, begin to have an admixture of strength with their richness. And while, again, association with the actual materials in which these textures generally occur has doubtless influenced our attitude toward them, that cannot be the sole determinant, for as far as cost goes they are all in the same class, yet each does have a distinctively different flavor of its own.

So definite are the qualities of these various surfaces that their use imposes certain restrictions on all the rest of the decorative scheme. The decided dignity of velvet precludes delicate furniture and demands in the decorations rich, rather deep colorings. The taffetas and moirés, being intimate, are appropriate only with a rather small scale, dainty type of furniture and are effective only in the higher color keys. Damask, again, can go with heavier styles, such as the Spanish and Italian Renaissance.

All these textures and many more are available in wallpapers. All the woven fabrics, from burlap to brocaded velvet, including various types of linens and cottons, are successfully reproduced. Plain and embossed leathers are well rendered and even certain forms of stone, such as Caen, are approximated. The great advance in wallpaper manufacture in the last few decades has been primarily an advance in rendering a variety of textures. Nor are these various papers imitations. There is no effort to disguise the fact that they are paper. They merely make use of interesting textures that do appear in other materials for the sake of the beauty and decorative utility of the texture itself.

That it is the texture rather than the imitation that is sought is evident in the number of papers equally carefully treated as to texture but which imitate nothing and are possible only in paper. There are, again, other papers in which the texture and the design are derived from different sources, as, for instance, a damask pattern on a paper that has, not a damask finish, but an embossing rather more like grass cloth. These combinations are made, because of the utility of that pattern and that texture, for a wide range of modern houses.

In many cases the texture of a wallpaper is an actual texture. That is to say, the surface of the paper really has the quality of finish of which it gives the effect. In a few cases this is due in whole or part to the kind of paper used, but more often it is ordinary paper and it is embossing in some particular pattern that gives the effect. There are, too, the flock papers that actually have the nap or silk surfaces in the patterns. But in some cases the effect is due entirely to printing, even such decisive surfaces as velvet being quite well approached by tricks of registering.

When the importance of texture is more generally recognized and more care is taken in the selection of textures in wall coverings, both in relation to its effect on color value, tone and scale and to its own intrinsic interest, interior design will have a substantiality and completeness that it now too often lacks.

CHAPTER IX

LINE AND SCALE IN WALLPAPER PATTERN

IN the selection of a patterned wallpaper, the two factors that must be taken into account in addition to the color and texture are the quality of the line in the pattern and the scale of the pattern. By using the right line quality the effect of the whole room can be enhanced. By using the wrong scale the room is given a handicap that not even the cleverest subsequent decoration can ever entirely overcome.

The fact that lines have quality is familiar to the draftsman, the etcher, the painter, even though they have not always formulated the knowledge that they have long used almost instinctively. And psychology has to some extent verified their experience and made it more specific. But the decorator has for the most part overlooked this valuable bit of information, though surely it is relevant to his problems not only in the selection of a patterned wall decoration but in the use of fabrics, in the assembling of decorative groups, in the choice of styles, even.

The vertical line is erect. It seems dignified, aloof, even, a bit restrained and constrained, or, if the suggestion is warranted by the occasion, spiritually soaring as in the Gothic cathedrals. A room with strongly emphasized vertical lines would seem very formal, rather remote. So a room of state is appropriately impressive with tall panels reaching up to a high ceiling as in the great drawing-rooms of the French Eighteenth Century palaces, where the loftiness of the ceiling is exaggerated by the smallness of the furniture so that the result is a very regal apartment.

The horizontal, on the other hand, is the line of rest and repose. The prairies seem utterly quiescent and resigned because of the inescapable flat horizontal line. A strong emphasis of the horizontal lines in a room serves to make it restful, to put it at ease. But too strong and too unrelieved a horizontality will make it seem heavy and dead, inert. So the once popular Craftsman furniture not only looked heavy; it looked lifeless.

Broken and curved lines both give the feeling of movement. The broken line seems, in proportion to the shortness of the sections, the sharpness of the angles and the variety of directions into which it scatters, more or less agitated, nervous, conflicting. A line jointed in long sections with obtuse angles following a fairly consistent pattern will seem to move more slowly, with more control and ease, while the line in little segments with many sharp angles zig-zagging here and there seems in an utter frenzy. The nervous irritation of some wallpaper patterns comes from the sharp call on the attention of such broken and angular line patterns.

The quality of movement in the curves varies similarly to that in the broken lines. The long, gradual flat curve moves slowly and with repose. The short, deep, heavy curve moves rapidly. But no curved line ever achieved the agitation that broken lines can express, for the curve even at its most exaggerated always has a regular, controlled movement. So if one needs the feeling of vitality without loss of poise in a pattern curves instead of broken lines should be used.

Any line running in a diagonal direction has some feeling of movement, also. It is clearly out of equilibrium and so in a state of transition. So no strongly diagonal pattern is restful. It always seems about to change and right itself, unless it is compensated by an equal emphasis on the opposite diagonal that seems to support it, as in a trellis where the balance is righted again by the equilibrium of the two opposite sets of diagonals.

A set of lines moving in different directions can always be arranged to compensate each other in this way and so the whole pattern can be brought to a balance by the skilful arrangement. But even a slight maladjustment of the lines in such a compensated pattern will throw the balance off again, leaving a tangle of confused lines struggling in all directions with no apparent cooperation or purpose. So a complex compensated pattern is the most difficult of all to design. And even when it is a complete success such a pattern will never have the complete repose of a simple horizontal design, for example, for the repose will be won out of struggle and the conflict and opposing tension will be operative and evident. But just by virtue of this such a pattern will have more vitality and strength than a simpler line arrangement.

Such are the general characters of the different classes of lines. Within these general classes lines and line combination can achieve a range of specific qualities that surpasses verbal designation. All of the moods and emotions with which we are familiar can be conveyed by the skilful and sensitive use of lines and many more shades and tones of feeling for which we have no terminology. Any of the great draftsmen in the history of painting will afford numberless examples that need for their full technical appreciation long and careful analysis, but perhaps the most striking and immediately moving examples of the powerful and expressive use of line comes in the art of the Far East—the painting of China and its descendant, the color prints of Japan.

Nor is this ascription of character to line merely sentimental imagination. The experience is both too intense with sensitive individuals, and too universal to be denied, and few miss the feeling altogether. Though an average person of limited artistic experience might not get the line feeling in a Sung painting he does get some inkling of it in a Gothic church, or in the prairies, or in the jagged silhouette of a mountain range.

Psychologists have found no final explanation for the ascription of specific feelings to different lines but they have suggested many possible reasons for the experience and perhaps the whole truth is in a combination of these various explanations. In the first place, in looking at a line we are ourselves really active and because we are absorbed in the line we attribute this activity directly to it. Our attention, our perception, that is, is active in following through the line, tracing its course, correlating its changes of direction into one impression and relating it to its surrounding lines. And this sharp activity of eyes and mind which differs for different kinds of lines we ascribe directly to the line.

But further than this, not only our eyes and attention and memory are actively engaged in perceiving a line, but our body takes a part too. For we have in our experience actually followed through some such a line as that we are perceiving, we have run along such a path or drawn such a curve, or thrown a ball with such a sweep or been knocked violently from side to side, as a broken line seems to be knocked, and in memory there is some echo of these experiences that floats vaguely to the surface and is associated with the line. As the psychologists say, we feel ourselves into the line, and with remembered and imagined bodily movements and attitudes go through the same course that it is following.

In some cases, probably many cases, this imagined bodily attitude that responds to the suggestion of a line is reenforced by actual bodily movements following the imagined ones. For we have always some slight bodily response to any mental activity, there is always an overflow of energy into the motor areas of the brain that stimulates some slight muscular reaction and this bodily echo is very apt to be influenced by the particular idea we are then interested in and, insofar as it is possible, imitate it. If the engrossing mental content that stimulates the movement is a line and we are already imaginatively following that line in attitude and direction, then the slight muscular movements will be sure to carry out further the line's rate and direction of movement. Thus we will have an inner imitation of the line of which we will not be directly conscious but which will make us feel more keenly the quality of that line's motion.

So the vertical line is dignified and soaring because we imagine ourselves in such a straight, uprising or rigid attitude and perhaps in a small way assume such an attitude. And the horizontal line is restful because, in part, that is the attitude of repose to which we are most accustomed when we lie down to rest, and perhaps in part, too, because as we imitate a horizontal line that is perfectly level it gives an even emphasis on both sides that puts us into more perfect equilibrium, and a nice balance is relaxing. Similarly the broken line is active and conflicting because in part it requires a constant rapid change of attention to grasp it and this involves a sharply accented effort and in part it is like our own experiences of being thrown from side to side and enduring uncertainty of direction. The curved lines have each the speed and direction that we have been used to experiencing in our own muscles when following such lines, and the diagonal is toppling because we know that no object can maintain a balance at such an unsupported angle. Thus the course of attention, previous experience, and actual movements of our bodies all combine variously to enable us to interpret and feel lines.

The importance of respecting the quality of lines in room design cannot be overemphasized yet it is often quite neglected. An informal room cannot be emphatically vertical and still retain its ease and relaxation. Many rapid curves or sharply broken lines are necessarily destructive of a restful effect. To define diagonal lines either intentionally or accidentally, as with a series of pictures hung in steps up a wall, is to create a feeling of confusion and impermanence. But, on the other hand, a room dead and heavy from too unbroken a horizontal accent can be relieved and enlivened by introducing some curved pattern, though it should not be too acute and nervous a curve or it would be completely out of relation to the rest of the room.

Similarly a room that wants to be gay and cheerful can obtain its effect only by playing upon curves and broken lines. So a room in very bright colors that are themselves intentionally gay should not in line contradict this character by being stiffly vertical or quietly horizontal. The quality of the line, in short, must be consistent with the purpose of the room and with all the other decorative features, and the lines must be chosen and composed with consideration for their qualities.

No less important in designing an interior

is the problem of scale. The relative sizes of the objects in a room must both be and appear consistent, for in certain cases the actual size and the apparent size may not coincide. Any object, for instance, that makes a decided demand on the attention seems larger, whether that demand be due to some quality of texture, some contrast of color, some marked outline or some particularly conspicuous pattern, than the same or a similar object in a less interesting form. And it is the apparent largeness of the first object that must be taken into account rather than its real dimensions.

If the furniture is large and heavy with large-sized ornaments it may somewhat relieve the effect if there are only a few pieces of it in a small room, but the appearance of being crowded will never be entirely avoided. The furniture will always look out of place and huddled because it is built on a different scale from that room. Conversely, to multiply the number of pieces of small furniture in a very large room will actually fill up the spaces but it will never make the room feel adequately furnished. Room and furniture and all of the decorative accessories, including all of the ornaments and patterns, must be designed in the same range of measurements.

The scale of a wallpaper pattern is especially marked because the unit of the pattern is repeated down the wall and so serves as a semi-consciously estimated unit of measurement. In proportion as the pattern is a good one and well managed we will be less conscious of the actual number of repeats both along the wall and up and down it, but even with the best and least obtrusive repetition we will have some vague appreciation of the number of times we can see it and that appreciation will operate as a kind of measure of the wall. At the same time it will act as a measure for the pieces of furniture for which that wall serves as background. So the size of the design on our wall coverings is all-important to the success of the room.

If the wall pattern is very large it will, of course, tend to make the room seem smaller. If it is very large so that there are a large number of repeats in succession it will make it look larger. There are, however, some exceptions to this rule, designs in which the repeat runs into itself so gradually, as in certain tapestry patterns, that it seems rather to stretch the wall in spite of its large size. With the more formal patterns, however, such as the damasks, the rule operates quite consistently.

The size of the wallpaper pattern can be used in helping to adjust to a small room furniture that is too large, or to a large room furniture that is too small. The problem is to select just that size unit that will make, in the first case, the room look as large as possible and the furniture as small, or, in the second case, the reverse. To make the room look large ordinarily the pattern selected should be rather small. But too small a pattern would make the furniture seem more enormous than ever. In the same way a large pattern that would help minimize the size of the furniture would make the room seem still smaller. So that the unit chosen must be of an intermediate size, and only experiment can determine for each room the exact proportion most successful. Similarly for a room too large for small furniture a design somewhere in the intermediate range must be used.

In addition to the importance of the size of the pattern in maintaining the scale of the room it has, too, some value in carrying out a consistent character. So a large pattern is pretty sure to be more dignified, a small pattern more intimate. So, too, a pattern of moderate size that looks larger because it is outlined with flat, slow curves will have greater dignity, while a larger pattern that is broken up into small subdivision will look more informal.

CHAPTER X

WALLPAPERS FOR PERIOD ROOMS

THE careful and consistent period room has passed in interior decoration except for the very wealthy collectors who wish to make their homes to some extent private museums. For the average householder the period room was at best only an imitation and an approximation and at worst it was a parody. Comfort and convenience were sure to take gradual precedence over exactness, so that after a little use the period room was very apt to have become a patched-up counterfeit of its own intention.

But though the exact period room has gone, we still must depend primarily on furniture of period designs. The great cabinet makers of the past have left us an heritage that we cannot afford to throw aside. They have at one time or another worked out most of the possibilities of furniture design in line, in woods and their finishes and combinations, and in ornament, so that an effort to get away from all historical precedents usually can result in nothing but the eccentric. Nor is there any reason for wanting to discard tradition, for the range and variety of designs available in the history of decoration assures an appropriate style for every need in every type of house and room. Modern life in all its functions can find a fit setting in some adaptation from one of the great periods of decoration. To be sure, there have been some recent designers, especially in France, who have builded on the old to make new forms and thus made real additions to the established patterns, but as yet their work is so little available for the general public that it ceases to be relevant in the average furnishing problem. To solve that we must rely on the good commercial reproduction of the established period types.

For the accessories of every period type our surest guide is the practice that prevails in the period itself. For the furniture is the product either of an individual genius, such as Reisner in France or the Adam Brothers in England, and that genius expressed itself with equal perfection of taste in the appropriate surroundings for their designs, or it is the natural outgrowth of a way of life and a kind of architecture, as is the case with Spanish and Italian furniture of the Renaissance and the succeeding century; and here again the original influence that shaped the furniture shaped all its accompaniments. So while we may and must adapt and recombine period precedents in modern decoration we had best find our key to the successful use of the established furniture types from contemporary custom.

SPANISH AND ITALIAN

Though much of the Spanish and Italian furniture of the Sixteenth and Seventeenth Centuries is too palatial in manner for the average modern home some of the simpler examples have a quiet dignity and directness that fill well the need of a living-room, library or dining-room that has ample space. Recognizing this, a number of manufacturers have made available reproductions of these styles. These simpler pieces succeed in having the sturdy straightforwardness that we sought in vain fifteen or twenty years ago in the passing fad of the Craftsman furniture.

The furniture of this period is practically all of oak toned a rich brown by time, a tone faithfully copied by antiquing methods in most of the reproductions. Originally heavy in design and ornament, this darkening of the wood has made this furniture even more heavy in appearance, so that it is necessary to have a rich, strong background for it. It is, moreover, very architectural in character, the ornament being deeply carved and built into it in decisive structural lines, so that the wall must emphasize the architectural structure of the room, not conceal it under too much applied pattern.

These demands are met by the wall finish most general in the old Spanish and Italian houses themselves, a smooth-finished plaster toned a rich light brown. But for a modern house this is an expensive and seldom successful treatment to reproduce, for the plaster must be delicately and subtly hand-finished to get the right surface quality; and the warm tone that age and weathering have produced in the original rooms with its delicate variations is extremely hard to approximate in paint. The usual mechanically hard and unmodulated plaster of our walls has no character and no strength whatever to support this rich furnishing, and the quality of brown that generally results from antique tan paint will conflict disastrously with the clear deep brown that should be the color of the furniture. The most effective wall that can be obtained for the usual room that is to be furnished in this type of furniture is, therefore, one of the wallpapers that have the essential qualities of the old plasters, a so-called engraved leather that has not too exaggerated modeling, either in a plain tan or, perhaps better still, in a blend that is not too mottled and multicolored. This does approximate the quality of the old plaster and gives a background that is definite and strong enough for the massive furniture. Traditionally it is appropriate for any room in the house from living-hall to bedroom, should that manner be attempted in the more personal rooms.

The one defect in such a wall covering for this furniture is that it is in danger of making the room seem dead and monotonous, too much drenched in tans. In the original palaces the effect was brightened with pieces of brilliant damask, gorgeous tapestries and colorful paintings. Where the minor decorations cannot be so varied it is advisable to get the relief in the wall covering itself. This can be done to some extent by using one of the blends that has more

Photo by M. E. Hewitt

THE CHINESE PAPERS ARE ENTIRELY IN KEEPING WITH CHIPPENDALE EVEN WHERE ONLY PART OF THE FURNITURE IS IN THE CHINESE MANNER.

Photo by M. E. Hewitt

AN ANTIQUE HAND PAINTED CHINESE BIRD AND FLOWER PAPER IS A BEAUTIFUL AND FITTING BACKGROUND FOR SHERATON.

color in it, especially one with dull red lights. Or it can be done and maintain an even more perfect period consistency by using a damask paper, one of the larger and conventionalized patterns taken from the damasks of the Sixteenth and Seventeenth Centuries in which fruits and flowers combine with formal patterns all in heavy curves. The two-toned damask is the most usual, and red and green, especially red, the most familiar colors in the really old examples; but the usual modern householder would hardly feel equal to sustaining this note, so he will probably have recourse to the cooler and flatter blues or even a tan which may seem sufficiently enlivened by the pattern. Similar possibilities, but for richer, more pretentious rooms, are the velvet brocade papers in large-scale patterns, either those that obtain their nap effect by the character of the printing or the real flock papers, and those papers that have taken their model from metal brocades so that there is a glint of gold or silver in the pattern. Akin to these latter, and frankly an adaptation, are some very useful papers with a grass cloth finish that have these same damask patterns printed either in a deeper tone of the same color or in a light metal line. These have the advantage of the strongly marked texture that helps support the decisive weight of the furniture.

Very characteristic, especially of the old Spanish houses, are leather-covered walls with heavy embossing and often rich polychrome enriched with metallic paints. These leathers have all been very successfully rendered in papers, not all, to be sure, fortunate in design but some of them really quite dignified and sumptuous. They make a rather somber room but for a library or even a dining-room where this effect is not amiss they are entirely successful in combination with Spanish or Italian furniture. Here again the designs should be large and in strong curves, flower groups and vases of flowers with the conventional settings.

The plain plaster wall or the damask or leather-covered wall is in accordance with the most familiar precedents, but there were variations on these and exceptions to them that might well be taken as a clue to lighter and more individual effects with the modern Italian furniture. There were, for instance, the various forms of frescoes which suggest patterned papers of similar designs. The all-over grotesque fresco of open design with the clearly marked architectural framework is well approximated in some of the all-over lightly drawn architectural patterns of papers, especially those in two tones. If the pattern is sufficiently strong and structural it certainly could sustain the furniture design, especially if the paper itself is of a firm heavy texture. A coat of varnish, not too heavy or too glossy, would aid further in harmonizing such a paper with this kind of furniture. An even lighter and gayer effect that still is appropriate takes its suggestions from the more colorful frescoes of Venice, rich with fruit and flower designs in red and gold and blue. But such a paper must be sharply distinct from a chintz rendering of these motives, more formal in drawing and flatter and more schematized in color.

Certain of the more severe printed linen designs in paper could be used in an informal Italian room, reminiscent of the painted linens still found here and there in Italian palaces, especially in Florence. A design in checkers or in strap work with conventional patterns, such as the fleur-de-lys, is what is needed.

And a delightful variant very much in character would be one of those occasional papers to be found with heraldic motives.

The closely derivative furniture—the Spanish Colonial with its bastard offspring, the Mission furniture—requires similar papers. They, too, can use the engraved leathers in antique tans that have the effect of the old plaster and the various blends. They, too, can use the heavy patterned damasks, but being much cruder the richer damasks are impossible, the velvets are out of the question, and all but the simplest leathers unwise. And, conversely, they permit of coarser types of papers. Those with the burlap textures, especially in the natural burlap color, are the most obviously fit. The grass cloths with their well emphasized texture are all available. Those papers that depend entirely on texture, such as the crêpes, and those with the granular plaster finish are exceedingly good. A tapestry paper, too, would not be unfit if not too luxurious in design.

When we turn from Spain and Italy and their derivatives to England we get a sequence of furniture styles that have been and must continue to be the main dependence of the average householders. England is, after all, the great land of the middle class and the great land of homes. So there has developed their kinds of furniture, useful and attractive and livable in the small utilitarian house. Especially in the Eighteenth Century the furniture design achieves a satisfactory practicality combined with quiet charm that cannot be surpassed for homelikeness.

WILLIAM AND MARY

The first distinctly English furniture that has general utility is that of the reign of William and Mary at the end of the Seventeenth and just the beginning of the Eighteenth Centuries. It is, in truth, an intensely English style, dignified, reserved, a little heavy, a bit somber. It is not, however, as massive as the Italian and Spanish types and it permits a much more varied range of wall decorations.

Here, too, of course, plaster is appropriate and so the plaster papers can be used, the engraved leathers, the rough plaster-finished types, and even the crêpe papers that give something of the unequal surface and rough strong play of shadows that troweled plaster has. Leather was at this time very much in vogue in all its forms, so the leather papers also are appropriate, whether the plain leathers, the embossed or the polychromed. Tapestry, also, was very much in favor. Quantities of it were being imported into England from Flanders and an English factory of considerable importance had been started, the Mortlake looms. So the verdure tapestry papers can well carry out the spirit of the time, especially those verdures that have taken their model from the Flemish Seventeenth and Eighteenth Century designs with a rather coarse heavy leafage in dull decided blue-greens, and large birds and bits of castle and garden vistas in tan. All these papers are, of course, rather more formal in their character, and the leather and tapestry types are quite somber. A little lighter in tone but still dignified and for the more formal rooms are the damasks of the period, the same damasks that are to be found in the Italian and Spanish houses, for England imported her silks and these were the two most important sources of her supply. So the same damask papers can be used with William and Mary as with the Italian, large heavily scrolled patterns in rich reds and blues and greens.

But there are more frivolous papers, too, that can be called on for the William and Mary rooms. For by this time wallpaper itself had begun to come into use and chintzes were quite general. The wallpapers for the period were primarily the hand-painted ones imported from China, and so the Chinese type of design should be used, not the small fantastic Chinoiserie of the later French period with Europeanized Chinamen but the designs really taken from the the paintings by the Chinese themselves, with the rather large but very delicately rendered flowers and birds or the sweeping decorative landscapes in vivid greens lighted with traces of cinnabar red and gilt, or those that show domestic scenes with solemn little doll people emerging from exquisite little houses.

The few European papers of the time that were in use in England are less useful as prototypes for modern decoration. They were the flocks, silk and wool, which are, of course, available to-day in brocaded velvet and damask patterns; the dominotiers from France with crude patterns in a printed outline, the colors painted in, designed usually in separate panels, or occasionally attempting a repeating continuous design; or the English printed papers in equally crude outlines and a few flat colors printed or similarly painted in. These latter types are really too crude to serve as models for our decoration.

The chintz of the time, however, is a very useful source of ideas for covering walls of William and Mary rooms. They were introduced into England during this reign and were received with great favor. Queen Mary herself had her bedroom at Windsor hung with chintz and the example was eagerly copied. These early chintzes had rather small, sparse patterns, thin flower sprays with linear, scrolling stems and small bright blossoms. The colors were dull reds and blues and yellows and greens in a limited number of tones and the ground was often écru, or, if it was not écru originally in the samples we still have time has toned it so that an écru ground chintz paper seems more fitting with the period, the more so that the soft yellowish tone is better adapted to the warm brown of the walnut furniture. But only chintz papers of this type, with the small patterns unpretentious in color and design, but delicate and varied, could be used with William and Mary. The large flowers heavily modeled

Photo by M. E. Hewitt

A CHARACTERISTIC LOUIS XV FRENCH WALLPAPER.

Photo by M. E. Hewitt

AN UNUSUAL ADAPTATION OF CHINESE MOTIVES MAKES A CHARACTERISTIC WALL DECORATION IN A LOUIS XVI ROOM.

and brilliantly colored, with gorgeous birds, that came into vogue a hundred years later would quite overwhelm William and Mary furniture.

Perhaps the most interesting and characteristic hangings of late Seventeenth Century England from which hints for wallpaper can be taken are the crewl embroideries. These wool embroideries, usually worked on cotton but sometimes on linen, were patterned after Indian silk embroideries that had been brought into England by the East India Company. The design follows one general precedent. A large tree with spreading branches runs the full height of the hanging, with great vari-colored blossoms drooping from the branches. The leaves are few in number, individually drawn, like the flowers. At the foot of the tree the earth is usually indicated in conventionalized mounds, sometimes with a row of small flowering plants, occasionally with a procession of quaint animals. The colors are dark green, dull blue, brown, usually in two tones, and dull Indian red. The traditional design has been reinterpreted with great success in modern commercial chintzes and could be rendered very well in wallpaper.

QUEEN ANNE

With the accession of Queen Anne a new character appeared in English furniture. Not, of course, that design was revolutionized instantly at the first moment when Anne sat down on the throne, but that a slowly growing tendency came into full development in her reign. This tendency was the trend away from the straight-lined, rather massive furniture, strictly structural in conception, that England had always used, to a lighter, more decorative type. The lightness was not achieved at once, for Anne brought with her her Dutch traditions to influence the fashions of the time, and native Dutch design is characteristically sturdy and sensible; but the feeling for the decorative line and decorative treatment became stronger and stronger, taking a lead from the French court that was dominating fashion, and from Oriental decorative arts that were fast flooding the market. So to the heavy Dutch forms that endured were applied curved lines, cabriolet legs, splat backs for chairs and benches with compound curves outlining the splats, bowed fronts to chests of drawers, carvings of scrolls and shells, all hints taken from France; and finally there was added red lacquer and black and applied decoration in gold, the Chinese contribution.

For the plainer types of furniture of the period that are more closely allied to the old Dutch-English tradition all the papers appropriate for William and Mary are still good, the large damasks and brocaded velvets, either in flat printing or in flock, as they had them at the time, the leathers and rich-toned tapestries and rough plaster simulations. The chintzes are still the same too—light little patterns in simple dull colors. And there can be added to these types any paper that takes its inspiration from gros-point and petit-point, for these were of greatest popularity at the time, a rather coarse heavy gros-point especially with scrolly leaves and thick flowers in dull reds and blues and a great deal of brown being especially approved.

But added to these established types were new fads of growing importance that offer suggestion for wall coverings. Especially well favored were the printed Indian cottons. The luxurious bedrooms of the period were time and again hung with these, with naturalistic designs of trees and flowers and animals printed in dull tones of the primary colors and of brown that were substantial without being somber. A paper taking its clue from these Indian cottons would be very delightful in a less formal Queen Anne room.

Above all, however, the Chinese painted papers and their modern renderings are appropriate for this Queen Anne furniture. The enthusiasm for Oriental decorations was at its height, wallpaper was well established in fashionable favor, the French producers had not yet achieved a finish and beauty of pattern sufficient to attract high-class patrons, so the Chinese paper had the field undisputed to itself.

Finally, there is a wallpaper hint to be taken from the Dutch ancestry of Anne. The blue and white tiles in Holland were not unfamiliar in England at this time, especially for facing the walls of the entrance hall. A courageous decorator could get a charming and typical effect by using one of the blue and white tile papers, especially one with richly pictorial tiles, that we usually associate with bathrooms, for the entrance hall of a Queen Anne cottage. But it may be asking too much of courage so to defy habit.

CHIPPENDALE

By the time Chippendale, whose shop was

producing from 1749–1779, rose to prominence the whole quality of decoration had undergone a decided change. The French and the Dutch influences that constantly struggled together during Anne's time resolved their contest and the French obliterated the Dutch. The Oriental strain, however, continued to appear in both design and decoration. And the French that predominated was, moreover, the French of Louis XV, gay, capricious, elaborate, extravagant. Chippendale was the disciple of the flowing line, the matter of complex curves, and he used them with an inexhaustible resource and fertility of imagination that bordered sometimes on the fantastic. But though Chippendale was under the sway of France, he never quite forgot that he was English and was designing for an English clientèle. So under the fanciful elaboration there were always strong, architecturally sound frames and in the delicate elaborateness of the decoration there was always restraint. Chippendale never duplicated the most fantastic and most extravagant flights of the French court designers.

The papers that were actually being made in France and used in both France and England at the time are obviously the most appropriate with this French phase of Chippendale, together with the modern papers adapted from the silk brocades of the period. Papers and brocades alike were most often in light fresh colors, the Boucher colors, that were particularly happy with the lightly constructed furniture. Dove gray was a great favorite; rose pink ran it a close second; Chippendale himself seems to have used corn yellow, and a pale blue recurs often in the fabrics of the period. The one basic difference in the colors for a Chippendale room and a Louis XV room is due to the difference in the wood finish predominant in the two styles. Louis XV furniture was all gilt or painted, usually in soft gray, and gilt or painted furniture can take fresher, brighter, warmer colors than the rich red of the mahogany that Chippendale used almost exclusively.

The patterns for these Louis XV Chippendale papers are the familiar ones of the period, stripes in graceful undulations scattered with tiny bouquets and bow-knots, stripes made of lace, arabesques and arabesqued vines suggesting stripes, the scroll and shell, and again ribbons in all forms and finishes as long as they are delicate, graceful and curving in outline. All of these are available for Chippendale rooms except the most extravagant, which hardly accord with his current of English restraint.

The Oriental influence that came into full power at this time in both France and England produced the highly individualized style of Chippendale Chinese. Whereas in the time of Anne Chinese lacquer decorations had merely been applied to the standard frames of the time and the imported novelties used with the real English furniture, now Chippendale took up these Eastern motives and made a new style of frame design that adapted the Chinese and the English in a new, exotic manner. And the accessories also that he used with the new Chino-English designs were usually Western interpretations of the Chinese: French Chinoiseri in which the real Chinamen would scarce have recognized themselves. Only occasionally did he depend on true Chinese pieces.

Papers that reproduced or reinterpreted these old Chinoiseries of such men as Pillemont, papers with trellises opening on bits of Chinese landscape, with scrolled frames enclosing fantastic trees and plump landscapes, with floral stripes alternating with stripes displaying slant-eyed beauties, are equally successful with the Chinese Chippendale and with the plain Chippendale or with a combination of the two. The old Chinese papers themselves or their modern prototypes are still useful too, but to keep in scale with the furniture these must be of the lighter more delicate types. The massive landscapes and strong floral patterns that support well the heavy furniture of Anne would rather overwhelm fragile Chippendale. The scenes of domestic life with their doll actors and the lighter flower and bird papers are the successful kinds. Then there are, too, for these forms of Chippendale some of the chintz papers, the chintz in lighter, smaller, flatter patterns or, even more exactly fit, the papers after the *toiles de Jouy,* with bits of rural scenes in strips or in verdure frames printed in monochrome, pale brown or gray on white.

There is, in addition to these two Chippendale types which merge and combine happily and can carry in general the same kinds of accessories, a rather more somber and pompous Chippendale made to meet another fad of the period, that for the Gothic. For Chippendale was an adept opportunist scenting ahead the popular trend, now to the French now back to the Chinese and now the Gothic, and meeting at every point the transient taste. Chippendale Gothic, which is far from Gothic, is of necessity built on heavier lines with pointed arches and trefoil silhouettes. Being heavier and a little more sober in spirit it takes a more solid and dignified pattern. Some of the richer, quieter brocades of the time are most appropriate.

ADAM

The ascendency of the Brothers Adam marks the reaction from the elaborate fantasies of Chippendale to the simple, the classical, the straight-lined. Gaiety was superseded by restraint, elaborate decoration by chaste refinement of proportion, extravagance by the real elegance of pure line and careful design. The Chinese motives gave way to the Roman as they had been interpreted by the Italian decorators, so that in place of lattices and pagodas there appeared urns and rams' heads. In short, the capricious, unintellectual feminine art that had been foisted on the world by Louis XV's mistresses was put aside for a more rational, coherent style of decoration, decisively but delicately rectilinear. For though this new style was classical it was still light and graceful. Mahogany continued to be the predominant material and mahogany lends itself to slender forms and delicate, exact construction. Slender forms in turn suggest light colors, so even though restrained almost to severity Adam decoration is still pitched in a high key.

The Adam Brothers themselves finished a great many of their walls in stucco, but they did use fabrics, also, so that fabric papers in the Adam style are entirely appropriate. Of these, the most usual were brocade and damask; neither the heavy, conventionally patterned damask that goes with William and Mary or Anne nor the elaborate flower and ribbon brocades fit for Louis XV and Chippendale, but damasks and brocades just designed for this furniture, with clear, open, straight-lined patterns in light and clear though not too weak and pale colors. Light-toned stripes are good and light-toned moirés, if they are of sufficiently good quality paper really to get the textile effect. The favorite Adam color was a peculiar shade of light green with a slight bluish cast. With this they usually used a pale flesh pink. Light grays they used too and occasional light blue.

In addition to these silk fabrics the Adelphi, as the Adam Brothers were known, sometimes used also tapestry covered walls, so that tapestry papers if light enough in design and color could be used with furniture of their type of design. Printed linen, too, was in general use in England at the time and designs similar in spirit were developed for it which have recently been adapted to wallpaper, classical grotesques either in back on white or in dull brownish red on white.

Besides these various fabric papers that are available and appropriate there are on the market a number of papers specifically of Adam design. Most of these show the familiar Adam motives, the urn, the ram's head, the oval plaque, and so on, usually printed *en grisaille,* that is in gray on gray to suggest stucco relief, but sometimes printed in two contrasted tones. A particularly effective type is printed in black and white on gray and uses the Wedgwood plaques that were so familiar in all Adam design. But though these papers are, of course, absolutely correct they cannot solve the wallpaper problem for all Adam rooms, partly because the Brothers Adam themselves used a wider range of fabrics, but primarily because they are too correct for any but the strictly period room.

HEPPELWHITE

The style of Heppelwhite is not entirely distinct from that of the Brothers Adam either in time or character. In fact Heppelwhite, who was a cabinet maker in contrast to the Adams who were architects, sometimes executed their furniture designs for them and his work extends over practically the same period. But Heppelwhite also made designs of his own, and though they are of the same general spirit as the Adam designs the true Heppelwhite pattern has an unmistakable individuality of its own.

Heppelwhite is like Adam in that it is light, elegant, exact, classical in inspiration and precise in execution. Like that of Adam, his furniture is primarily in mahogany. But whereas Adam is emphatically rectangular Heppelwhite liked to vary his line with long flat curves and flattened ovals. So Heppelwhite came closer to the true Louis XVI from which, in fact, he took many suggestions. But he was always an English Louis XVI just as Chippendale was an

English Louis XV, not quite so elegant, so elaborate, so aristocratic as the French, rather more usable and homelike. And he was rather more usable and homelike than Adam, too, more intimate and practical for everyday use.

So while the wall coverings to go with Heppelwhite furniture must be light and rather elegant they can be, also, more informal than those for the pure Adam room. The same light-toned and delicately but rather severely patterned brocade and damask papers, the light-toned striped fabrics and the same light tapestries that are appropriate with Adam designs are also good with Heppelwhite. But in addition to these there is a range of chintz and printed linen designs that are rather too informal for Adam but are quite in keeping with Heppelwhite. These chintzes are those with more elaborate floral designs that came into fashion in the middle of the Eighteenth Century. While they are not the most elaborate and rich chintz patterns, not those with huge roses and peonies and gorgeous peacocks or tropical birds that came into use a few decades later, they are much richer than the sparse field flower patterns of William and Mary and

Anne, having medium-sized rose patterns, vines and small flower sprays that are more varied and thickly strewn than in the early pieces. Particularly usable are the striped papers with flower garlands on the alternate stripes, either in a silk or in a chintz finish.

SHERATON

Sheraton is another style practically contemporaneous with Adam and akin to it but still maintaining its own character. Like Heppelwhite Sheraton was a cabinet maker. His work is a bit later than that of the Brothers Adam and Heppelwhite—he produced designs into the opening years of the Nineteenth Century—but most of it is very similar in character to theirs. It is straight-lined, light, precise, restrained. But it has a quality of directness that is more friendly than the rather aloof Adam, a bit more personal than Heppelwhite, even.

Like Heppelwhite, Sheraton can take various kinds of silk papers. Sheraton in his design book speaks of using for upholstering plain silks, striped silks, flowered silks and gold and silver brocades; and all of these in paper reproductions, except possibly the too sumptuous metal brocades, would be appropriate for the wall covering. The color to which Sheraton makes specially emphatic reference is blue, and certainly it is a successful color with mahogany. Sheraton speaks, too, of the three pleasant combinations with blue—blue and white, blue and black, and very pale blue and yellow—good hints all of them for Sheraton rooms. The lighter tapestries also can be used with Sheraton furniture and all of the chintzes and toiles that go well with Heppelwhite.

But in addition to these Sheraton, being somewhat later and a little stronger in construction, can take those bolder, richer chintz designs that have been so beautifully adapted to wall-paper. The heavy, leafy vines with the huge exotic blooms, the sweeping-tailed peacocks, the brilliant macaws, all go well with the strong simplicity of Sheraton. Other very attractive chintzes that have come down to us from that time and been reproduced in papers are those designed in panels, each panel a complete unit with a rich full basket of flowers or of a fruit and a clearly defined border. It was at this time that chintzes on a colored ground, also, became popular and these are most attractive in modern paper renderings.

Sheraton himself used wallpaper extensively even in his most luxurious apartments. But he made it more decorative and more architectural, usually, by applying it in carefully designed panels with multiple and varied borders. In Part I of his Cabinet Maker's and Upholsterer's Drawing Book Sheraton shows a drawing-room with small panels between the windows and over the fireplace and alternating with larger panels on the unbroken walls. Each of these narrow panels is defined by an architectural border, very narrow, suggesting columns and a molding. Above each of these is a panel showing a classical scene in grisaille, suggestive of a stucco relief. The large panels run clear to the frieze, which is indicated by a narrow guilloche border surmounted by another wider one in the classic honeysuckle design, and are outlined with moderately wide borders edged at either side with narrow ones.

In Part II he shows the south end of the Prince of Wales drawing-room which is equally complex in its paneling. Above a dado there are wide panels in a fine vertically striped paper defined by a narrow cut-out architectural border, the whole mounted on the same stripe used horizontally so that this horizontal stripe shows

Photo by M. E. Hewitt

THE EMPIRE IS THE PERIOD OF THE GREATEST VOGUE OF LANDSCAPE PAPER. AN OLD EXAMPLE.

as a moderately wide border on all sides of the architecturally indicated panel. Narrow panels, between the wide ones and between doors and windows, have Chinese scenes simply framed in a narrow border and also mounted in the same way on the horizontallly striped background. Subordinated panels are used to flank the wide panels on very long unbroken spaces, these being in a plain conventional lattice design. The whole series of panels is topped with a Greek frieze in the honeysuckle design, and the ceiling also is striped with the same fine stripe and has a narrow geometrical border about a foot from the cornice all around. So Sheraton made complex but careful combinations of his papers and his borders to give them a stately or elaborate decorative effect.

But perhaps the finest and most characteristic paper for a Sheraton room is one of the landscape papers that were then produced in France, not so much those of Chinese scenes or landscapes as the travel papers and the illustrations of romances. These in their vibrant flat-printed colors with their quaint and varied scenes are altogether charming with the rather plain Sheraton furniture, especially when used, as they most often were at the time, about a white painted dado. For the Sheraton dining-room there is nothing to take precedence over these landscape effects that are still being printed to-day, in many cases from the original blocks.

When we turn from England to the contemporary styles in France we have types of furniture less generally useful. For France in the Eighteenth Century was entirely dominated by her court. Her decoration was the sumptuous decoration of a particularly extravagant royalty. Even when it was adapted to the middle class it remained ornate beyond the habit of most American homes.

LOUIS XIV

The furniture of Louis XIV is of no use in the average household. It is a style for the very wealthy, demanding luxurious surroundings and elaborate service. It is intolerable except in the original pieces which, when they are good, are worth each one a fortune. The reproductions that have been attempted here and abroad are tawdry and ostentatious, too distressingly imitations. Any one who has these valuable originals can and will set them in the appropriate surroundings of rich brocades or heavy and sumptuously carved panelings. So there is no question of wallpaper for Louis XIV.

LOUIS XV

The styles of Louis XV and XVI, however, do have a somewhat more general use. They were rendered in simpler forms and materials and their light frames and dainty proportions have made them quite popular for women's rooms. They are, in truth, women's styles, made for and determined by the beauties of the two extravagant and pleasure-seeking kings. Louis XVI, rather more severe in its mock classicism is useful, too, for small drawing-rooms and dining-rooms, especially in modern apartments where heavier furniture leaves no room for human occupants.

The paper for the Louis XV room is the same as that for Chippendale, brocades and damasks in the pale colors of spring flowers, jonquil yellow, lilac, rose and light blue and in gracefully curved, light patterns on a small scale, undulating stripes, delicate trailing vines, bow-knots, bunches of flowers, scrolls and shells and Chinoiseries, both the French fantastic Chinoiseries, that at this time were at their height and the real imported Chinese papers. *Toile de Jouy* papers, also, are characteristic of this period, and they, too, show Chinoiseries or bits of landscape framed in trees and flowers.

LOUIS XVI

For the furniture of Louis XVI the papers are the same as for Heppelwhite, still light but a little more restrained and classical than in the preceding reign, with straight stripes instead of undulating, classic motives rather than so much of the Oriental, though Chinese paper was still in use, and with pastoral scenes especially for the *toile de Jouy* papers in honor of Marie Antoinette's play at rural life. Satin was particularly fashionable for room decoration at this extravagant time, and silks with satin stripes, so satin paper or satin-striped paper is very appropriate.

DIRECTOIRE AND EMPIRE

The styles of the Directorate and the Empire, though contemporaneous with Sheraton, for the most part, are rather different in character. Because they do fall in the same period when the same kinds of wallpaper were being made they do have certain styles in common. The many-block printed landscapes, for instance, are equally appropriate for Directoire and Empire furniture and for Sheraton, and the same richly patterned chintzes and chintzes on colored grounds are good for both. But in addition to these, the French styles have certain papers peculiarly appropriate to them. There are, for instance, all the old papers and the old *toiles de Jouy* that have been reproduced in papers that show republican symbols that go very well with the Directoire; and for the Empire there are the papers showing episodes of Napoleon's life, Egyptian designs of the pyramids and the sphinx and the lotus reminiscent of Napoleon's Egyptian campaign, and the more emphatically classical papers with Roman grotesques or Greek shell patterns that flattered Napoleon's assumption that his Empire was a reincarnation, as it were, of the great Roman Empire. Especially Empire furniture is much heavier than the contemporary English styles and requires a heavier type of paper, larger in scale, with deeper, richer colors. Typical favorite colors of the time were chocolate brown, especially mottled with a combination of rose and green, and purple, reminiscent, of course, of the imperial purples. Purple would be almost too overwhelming for modern taste, but the mottled chocolate brown can be successfully repeated in the soft Tiffany blends with their warm tans merging into traces of rose and green.

PAINTED PEASANT STYLE

From one other source besides the regular period styles modern furniture has borrowed designs—from the painted peasant furniture of various parts of Europe, particularly the Tyrol. This charming furniture, with simple sturdy forms, gay colors and smart applied decorations, is available for any of the informal rooms. It presents a special problem in wallpaper selection, for its own bright colors are in danger of clashing with equally gay coloring on the wall, yet its vividness forbids flat colorlessness in the background. In the old Tyrolean houses where it originated the walls were either wood-paneled or plastered with the softly modulated hand-finished plaster in warm cream tones. The plaster papers and those of similar texture, the engraved leathers and the crêpes are therefore very useful with it. A coarse stipple paper, too, goes well with it, for it has strength of finish enough to sustain the sturdy construction of the furniture and the stippling gives an opportunity for a little warmth of color. Grass cloth, too, while it certainly is unknown to the peasants of the Tyrols, nevertheless in both its silver and its gold shades makes a delightfully fit background for their furniture. And finally, if the decorator has courage, many of the chintz papers can be used to make a flaringly flower-like room.

CHAPTER XI

WALLPAPERS FOR OTHER ROOMS

THE living-room in the ordinary American home seldom makes any pretense of being even a partially consistent period room. Because, usually, it combines the three functions of drawing-room, sitting-room and library, and because it has been shaped generally by the tastes and needs of the whole family, it is quite apt to contain a widely assorted combination of furnishings—a few wedding presents, some pieces donated by relatives, some overstuffed for comfort and some wicker to fill in, plenty of books on open shelves and all kinds of odds and ends of ornaments. Such heterogeneousness makes it unavoidably informal but, on the other hand, it is usually the only reception room in the house and so at the same time it must preserve some aspect of dignity.

To find a wallpaper suitable for a conglomeration of furniture and decorations that will

Photo by M. E. Hewitt

GAY AND AMUSING DESIGNS THAT WOULD BE TOO OBTRUSIVE ELSEWHERE CAN FIND A HAPPY PLACE IN THE KITCHEN

still have some slight suggestion of reserve and formality is not easy. Small wonder that the usual book on decoration takes the straightest road out of the problem by recommending a plain wall in either one of the inevitable neutral colors, tan or gray. They are entirely safe but porportionately uninteresting. There are, however, modifications of the plain wall that gain real quality through a decided texture Such, for example, are the crêpe papers, especially attractive in the heavy grades with wide uneven folds; the engraved papers when the modeling is not too exaggerated and spotty; in cheaper papers the oatmeals, certain odd papers with a sort of rough plaster finish; and the heavier cartridge papers. The interest of the surface finish adds to the otherwise characterless neutral wall and, at the same time, gives a strength to the room that is absolutely essential if there is any heavy furniture or many shelves of books.

A little more decided in quality than these papers but still within the safe prescription of plain neutral colors are the papers with texture so strongly defined that the texture almost constitutes a self-toned pattern. Such for instance are all the grass cloths, hardly more than a pleasantly roughened surface from one point of view, from another a simple but varied pattern in horizontal lines. The grass cloth papers, fairly successful imitations of their model, come within this class, too, as do the burlap papers that reproduce the various weavings of sack-cloth.

The next step away from monotony but still entirely within the bounds of safety are the blends. Blends, of course, are of many kinds and achieve many different levels of success. At the best, whether it is the simple blend in two tones of one color—for example, two shades of tan floating into each other—or the more elaborate kind introducing several different colors into a moving variation of tone, the blend is a nice suggestion of light and shade to relieve the dullness of a flat color. But at the worst the blend looks like a tubful of colored calicoes, all run into each other. Good blends are especially good in the engraved papers, and they have been managed well, too, in a rather more linear effect, in some of the grass cloth papers.

A type of paper that has a little more interest than the plain color or blend and still is not obtrusive is the two-toned damask pattern.

This has the quietness necessary in the background of a heterogeneous decoration and the touch of dignity that is desirable. These old textile patterns, gray on gray or tan on tan, give a well adapted touch of richness without being in the least pretentious. Especially attractive are such patterns in grass cloths or grass cloth finished papers. For a little more vitality and interest there are these same cloths in damask patterns with the pattern lightly outlined in metal silver or gold.

Another variation on the plain color that can be used with the lighter, chintz hung types of living-rooms is the self-color stripe, with the stripe indicated only by a different texture. If the room is very informal—the living-room, say, of a country cottage or a New England farmhouse sort of home—the stripe might be one of those that are lightly dotted in a contrasting color, faintly reminiscent of cotton prints.

For a very rich living-room background that still will not be very decided, there are the tapestry papers. These range all the way from the vague verdure effects, that are hardly more than massed lights and shadows in two or three tones of the color, to the rather clearly drawn landscapes; and they come not only in the conventional greens and green-blues but in soft grays and browns as well. Especially interesting is a recently issued English paper of which the design is adapted from a French Gothic hunting tapestry with medieval gentlemen pursuing quaint beasts through a thick and flowery wood indicated in the *mille fleurs* manner in flatly drawn, superimposed shrubs and flowers. It is rendered in soft dark colors and so is not too insistent.

Where there is opportunity for more decided pattern the closely drawn, continuous designs such as those which William Morris made so beautifully are very good. Ideal examples are his Pimpernel and Honeysuckle, both very fine in the right kind of living-room; but there are, too, modern American papers in the same manner that would succeed. Because of the shadings and variations in color these all-over designs permit of a richness in color, soft but strong blues and greens, that would be intolerable in a flat unbroken wall.

All-over designs of architectural ornament, are also available in color combinations that are not also emphatic and they range in tone and scale all the way from the lightest to the strongest, so that they can be fitted to any size of room and any kind of furniture. Printed in Cameo effect they are especially attractive and the grounds are in soft but clean colors that can set the color note of the entire room.

And finally, for the living-room that is quite English and rather more like a morning-room and very gay there are the chintz papers. But these forbid much ornament through the rest of the room. They do not accord with massive rows of heavy books, and they are too informal for any but a country house or a house in which there are other rooms for receiving guests.

THE DORMER WINDOW ROOM

Another special problem in the choice of wallpaper arises when there are rooms with dormer windows, and with the revived interest in the Dutch Colonial architecture and in remodeling old houses it has become a rather common problem. The many angles and broken surfaces of such a room make certain patterns impossible, the number of hidden corners in which there are apt to lurk dingy shadows make high keyed colors important, and the lowness of the ceiling also makes certain demands on the design. Plain colors are, of course, the simplest solution, but the charm of these rooms is in their old-fashioned quaintness and it can be realized to the full only with a quaint, old-fashioned patterned paper. Of these the most successful are the all-over patterns in rather small scale, floral patterns, preferably rather closely covered. But the floral pattern cannot be on a trellis background, for the many angles would make the trellis look as if it had been wrenched by an earthquake; nor can there be birds in it to have their backs broken at inconvenient corners, nor, for the same reason, human figures. The Chinoiseries are therefore out of the question. Only the plain continuous floral patterns succeed and these are best when there is a decided upward trend to the design to make the low ceiling seem a little less oppressive.

No paper in which the design stands apart as a complete unit can be used very well, for wherever the unit was broken by an angle of the wall effect would be lost, and as this happens frequently with dormer windows the resultant appearances would be chaos. So all the *toile de Jouy* patterns that have little rural scenes framed in branches or groups arranged at intervals on stripes are taboo.

But, contrary to expectations, plain stripes are entirely successful in a room like this. Their straight lines adapt themselves easily to the changing angles of the wall, a break in the unit is not disastrous, and the strong vertical effect helps lift up the ceiling.

KITCHEN

Conventionally, only tile papers are fit for a kitchen. Fortunately in recent years these tile papers have been designed with a little more respect for their æsthetic effect, so that one can make with them now quite a charming kitchen in blue and white in the Dutch effect. They have the advantage of being finished with a gloss surface that is impervious to steam and can readily be cleaned.

But other papers, too, can be made equally steam resisting and cleanable with a coat of varnish, so that the kitchen need by no means be limited to regular kitchen effects. After all, a kitchen is a great opportunity for amusing experiments, for we need not take it as seriously as we must a living- or dining-room and we can enjoy there startling papers that would soon be a bore in the more conventional rooms. Nor is the decorative problem of the kitchen negligible, for as the world stands now a large part of the American women have to spend many hours a day there, weary, stupid hours that need the cheering of bright colors and capricious patterns.

Somehow an out-of-door effect seems always welcome in a kitchen, so that chintz papers with bright flowers are very fit. Especially can one get a happy garden effect with some of the trellis papers. There is amusing appropriateness, for instance, in fruit trees on a trellis, and lovely color, especially in an orange pattern, that shows the bright ripe fruit and the dark green leaves on a lighter green trellis. The modernist papers, too, offer some charming fruit groups, and the kitchen is a good place to try out something that seems a bit bizarre. Then there are the Chinoiseries that are too colorful and too distracting for a soberer room. And even an informal landscape or pictorial paper might give a pleasant widening to the restricted domestic horizon. After all, charm need not conflict with efficiency but in the final analysis should enhance it. The kitchen because it is given to dull labor need not be dull. Rather it can the better bear the compensation of all those papers that are too strange, too vivid, or too striking to be attempted elsewhere yet on the shop rack look too delightful to pass by.

THE END

WALLPAPER DESIGNERS

WALLPAPER DESIGNERS

The list, as far as possible in chronological order, stops at 1830, the time of the introduction of machinery. All the firms for which each designer worked are given as far as is known.

1700	Blandin	*Adam*
17—	Blondel	*Le Sueur*
	Vincent Pesant	*Le Sueur*
	Panseron	*Le Sueur*
	Jean Pillement	*Le Sueur*
177-	J. François van Dall	*Reveillon*
	Percier	*Reveillon, Jacquemart and Bénard*
	Salembier	*Reveillon*
	Joseph Laurent	*Reveillon*
	Mèry père	*Reveillon, Le Grand*

WALLPAPER:

	HUET	*Reveillon, Jacquemart and Bénard*
	J. J. FAN	*Reveillon*
	CIETTI	*Reveillon*
	PRIEUR	*Reveillon, Jacquemart and Bénard*
	PAGET	
	LAVALLÉ-POUSSIN	
	CAUVET	
	ROUSSEAU FRÈRES	
	CHARLES MONNET	
	JOSEPH SAUVAGE	
	BOISSELLIER	
	FONTAINE	*Jacquemart and Bénard*
	ANDOUIN	*Le Grand*
	LEGENDRE	*Le Grand*
	PRUD'HON	*Bellanger and Dugourc*
18—	MONGRIE	*Zuber*
	RUGENDAS	*Zuber*
	HERMANN	*Zuber*
	CHABAL	*Zuber*

Dussurguey	*Zuber*
Dumont	*Zuber*
Laurent Malaine	*Zuber*
Mongin	*Zuber*
Ehrmann	*Zuber*
Zipelius	*Zuber*
Fuch	*Zuber*
Laffitte	*Dufour*
Madèr père	*Dufour, LeRoy*
Mader fils	*Dufour*
Wagner	*Dufour*
Portelet	*Dufour, Dauptain*
Dèlicourt	*Dufour*
Fragonard fils	*Dufour*
Alexandre Evariste	*Dufour*
Brocq	*Leroy*
Martin Polich	*Dauptain*
Aimé Chenevard	*Dauptain*
Martin	*Dèlicourt*
Wagner	*Dèlicourt*
Riesner	*Dèlicourt*
Dumont	*Dèlicourt*

WALLPAPER:

CH.-L. MÜLLER — *Dèlicourt*

THOMAS ROWLANDSON

MERKE

WALLPAPER PRINTERS AND DEALERS

WALLPAPER PRINTERS AND DEALERS

The list, which stops at 1830, is as far as possible chronological, the dates being those at which the firms appear in some records to indicate the general period of activity. When the address is known it is noted.

1568	HERMAN SCHINKEL	*Delft*
1620	LE FRANÇOIS	*Rouen*
1634	JEREMY LANYER	*London*
1670	TIERCE	*Rouen*
1670	JOHANN HAUNTZSCH	*Nuremberg*
1691	WILLIAM BAYLEY	*England*
1688	JEAN PAPILLON	*Paris*
1700	ADAM	*Paris*
17—	LE SUEUR	*Paris*
1727	ROUMIER	*rue Jacob-Saint-Germain, Paris*

WALLPAPER:

	DUFOURCROY	*Paris*
	MASSON	*Paris*
	MIYER	*Paris, (Successor to Masson.)*
	BASSET	*Paris*
	FORCOY	*Paris*
	VASEAU	*Paris*
	GOUPY	*Paris*
	BRETON	*Paris*
	LETOURNY	*rue d'Orléans, Paris*
	RABIER	*Orléans*
1730	BOULARD	*Orléans*
1735	M. N. B. POILLY	*Paris*
1740	LANGLOIS VEUVE	*Paris, (Successor to Papillon.)*
1746	JOHN BAPTIST JACKSON	*Battersea*
1750	GEORGE AND FREDERICK ECKHARDT	*Chelsea*
1750	JACQUES GABRIEL HUQUIER	*Paris*
	JACQUES CHAUVAN	*Paris*

1754	THOMAS VINCENT	*Fleet Street, London*
	ROGUIÉ	*rue Cloître-Saint-Germain, Paris*
	DIDIER AUBERT	*rue Saint Jacques, Paris. (For a time used Papillon mark.)*
1755	MASEFIELD	*Strand, London*
	MATT DARLEY	*Strand, London*
176–	LECOMTE	*Lyons*
1762	GARNIER	*rue Quincampoix, Paris*
1766	LANGLOIS	*Paris. (Successor to Langlois veuve.)*
1768	ECCARD	*Paris*
1769	JACQUES CHEREAU	*Paris*
	NIODOT	*Place du Louvre, Paris*
	LANCAKE	*Carrière (Manufactory)*
		rue Geoffrey Lanier, Paris (Shop)
1770	PANCET	*Paris*
	DAUMONT	*Paris*

	CRÈPY THE ELDER	*rue Sainte Appolline, Paris*
	WATON	*rue Sainte Appolline, Paris*
1774	DEMOISELLE HENNERY	*rue Countess d'Artois, Paris*
177–	REVEILLON	*Faubourg Saint Antoine, Paris. (Manufactory.)* *rue Carouzel Paris. (Shop.)*
1777	MATHON	*Paris*
1779	WINDSOR	*rue du Petit Vauguard, Paris*
	DANIER	*rue Dauphine, later rue de Bussey, Paris*
1781	DE COUVIER	*Bavaria*
1781	ARTHUR AND ROBERT	*rue Louis le Grand, Paris*
1783	JACQUEMART AND BÉNARD	*(Successors to Reveillon.)*
1788	CHOUARD	*Lyons*

178-	LE GRAND	*rue d'Orléans, Paris.* (*Manufactory.*) *Place Dauphine, Paris.* (*Shop.*)
1800	JOSEPH DUFOUR	
	LE ROY	*rue* 9½ *Beauveau, Paris.* (*Successor to Dufour.*)
	DAUPTAIN	*rue Blanche, Mibray*
1810	SEMON	*Jardin des Capucines, Mibray.*
	SEMON AND CORTULOT	(*Successors to Semon.*)
1821	MADÈR	
18—	DAMIENS	*rue de Bussy, Paris*
	MONTRILLE	*rue Vivrenne, Paris*
	JACQUES ALBERT	15 *rue du Bac, Paris*
	DAGUET AND CAFFÈRE	4 *Place Vendôme, Paris*
	LEGENDRE	*rue Pâte-Sainte-Antoine, Paris*
	CARTULAT	*rue Napoléon, Paris*

	PAULOT AND CARRÉ	*5 rue Reuilly, Paris*
	FRESNARD FRÈRES	*Paris*
	PERIGUEUX	*Paris*
	VITRY	*Paris*
	MASSON AND CHICANEAU	*Paris*
	VÈLAY	*10 rue Lelois, Paris*
	VAUCHELET	*Paris*
	BOULANGER	*rue St. Benôit, Paris*
	J. GUILLOT	
1823	SPRÖLIN	*Vienna*
	SCHÖPPLER AND HARTMANN	*Augsburg*
1825	HENRY WILLIAM	*rue Charleton, Paris*
	BOURIER	*Besançon*
	LANGIER	*Nancy*
	CANOLIS	*Nancy*
	LE FLAGUAIS	*Caen*
	PIGUET	*Lyons*
	RICHON	*Saint Genis*
	MOGNAT-PERRIN	*Vienna*
	WÈRY	*Vienna*

183–	DÈLICOURT, later and Campas and Gurat	
	BUZIN,	(*Successor to Dèlicourt et al.*)
	DEFOSSÉ, later and Karth	(*Successor to Madèr.*)

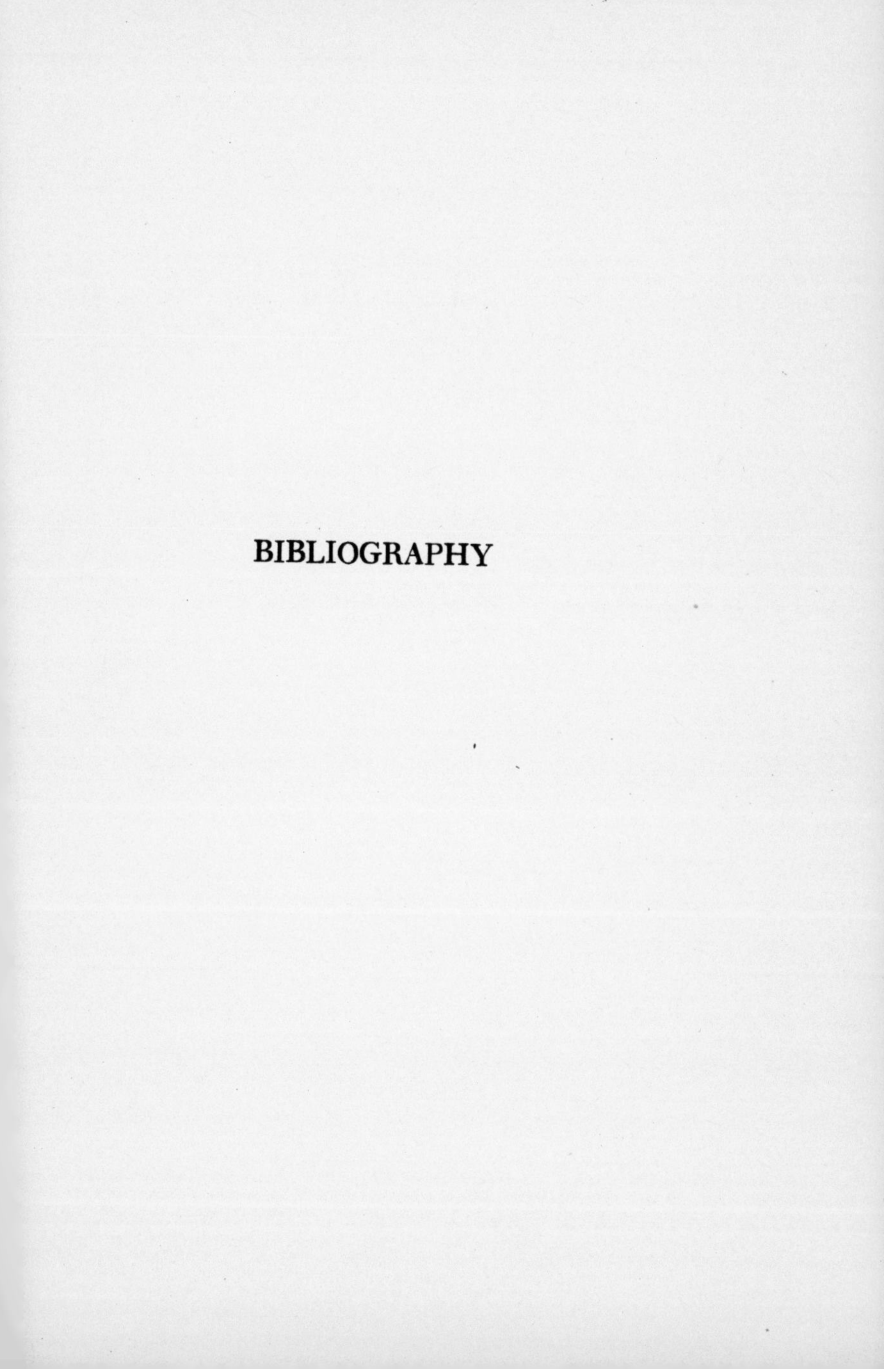

BIBLIOGRAPHY

BIBLIOGRAPHY

There is no attempt to make this bibliography complete. Many of the magazine articles on the subject are so without permanent value they are not worth listing. Some of the references cited are valuable primarily for the illustrations.

BRADSHAW, W. R. *Wallpaper, its History, Manufacture and Decorative Importance. New York, 1891.*

Some facts on history in America.

CRACE *History of Paper Hangings.*

FISCHBACK *Beitrag zur Geschichte der Tapeten Industrie. Darmstadt, 1889.*

FOLEY *Book of Decorative Furniture, Vol. I. London, 1910.*

Incidental historical facts.

FOLLOT, F. *Rapport du Comité d'Installation de la Classe. 68, Papiers Peints a l'Exposition Universelle. Paris, 1900.*
Very valuable source book.

GUSMAN, PIERRE *Panneaux Decoratifs et Tentures Murales de XVIII Siècle. Paris.*
Facts derived from Follot.

HAVARD, H. *Dictionnaire de l'Ameublement, Vol. IV, under Papiers Peints.*
Most complete account French history.

HESSLING, EGON *Le Style Directoire, Etoffes et Papiers. Paris.*
Some historical facts of this period.

HUNTER, G. L. *Decorative Textiles. Grand Rapids. Mich. 1918.*
Scattered historical facts and some remarks on design.

JACKSON, JOHN BAPTIST *An Essay on the Invention of Engraving.*
One of most important source books.

LENYGON, FRANCIS *Decoration in England from 1660-1770, London, 1914.*
Interesting incidental facts.

NORTHEND, MARY *Colonial Homes. Boston, 1912.*
Some descriptions early French papers in New England houses.

PAPILLON, J.-M. *Historique et Practique de la Gravure sur Bois. Paris, 1776.*
One of principal source books.

PERCIVAL, MC-IVER — *Old English Furniture and its Surroundings from the Restoration to the Regency. London, 1920.*
Incidental facts.

SANBORN, KATE — *Old Time Wall Papers. New York, 1905.*
Random facts. Excellent illustrations.

SEEMAN, THEDOR — *Die Tapete. Vienna, 1882.*
Some historical data.

MAGAZINE ARTICLES

ARTS AND DECORATION — *Vol. 10, pg. 260 S. W. Woodruff, The Romantic Story of Wallpaper.*

CONNOISSEUR — *Vol. 47, pg. 79 McIver Percival, Old Wallpapers.*

Vol. 52, pg. 83 Oliver Brackett, English Wallpapers of the Eighteenth Century.

Vol. 62 pg. 25 McIver Percival, Jackson of Battersea and his Wallpapers.

Vol. 62, pg. 156 M. Jourdaire, Some Early Printed Papers.

DEKORATIVE KUNST — *Vol. 20, pg. 573 Paul Westheim, Das Ornament an der Wand.*

Vol. 22, pg. 246 Neue Tapetenmuster.

DEUTSCHE KUNST UND DEKORATION — *Vol. 21, pg. 363 Franz Servaes, Neue Deutsche Tapeten.*

GAZETTE DES BEAUX ARTS — *Vol. 54, pg. 131 Henri Clouzot, La Tradition du Papier Peint en France.*

WALLPAPER:

Vol. 56, pg. 42 Henri Clouzot, Papiers Peints de l'Epoque Napoleonienne.

GOOD FURNITURE *Vol. 15, pg. 67 Selling Wallpaper to Harmonize with Other Decorative Fitments.*

Vol. 16, pg. 295 George Leland Hunter, Ancestors of Wallpaper.

HOUSE BEAUTIFUL *1914 pg. 153 Ann Wentworth, Decorative Wallpapers.*

1915-16 pg. 148 Phil Riley, and Frank Cousins, Landscape Wallpapers.

1921 pg. 369 Edward B. Allen, Ye Old Picture Wallpapers.

HOUSE AND GARDEN *Vol. 39, pg. 23 R. R. Goodnow, A Cinderella Room.*

Vol. 40, pg. 46 The Dispute over Wallpapers.

L'ART DECORATIF *Vol. 27, pg. 117 Ferdinand Roches, Vieux Papiers Peints.*

Vol. 28, pg. 175 L. Bouteille, Les Vieux Décors en Papier Peint.

INTERNATIONAL STUDIO *Vol. 75 pg. 48 Mary Northend, Wallpaper of the Olden Time.*

INDEX

Adam, 24
Adam green, 135
Albert, Jacques, 54
Andouin, 49
Arthur, 47, 48
Aubert, Didier, 26, 40

Banks, Sir Joseph, 16
Basset, 25
Bayley, William, 21
Bellanger, 49
Blondel, 24
Boissellier, 44, 45
Borden Hall, 5, 42
Boucher, 48
Boucher fils, 48
Boulanger, 55
Boulu, Charden and Carnes, 68
Bourier, 55
Breton, 25
Brocq, 48, 52
Bumstead, 59, 69
Buzin, 53

Campas, 53
Carrière, 41
Cartulat, 54
Cauvet, 44
Chabel, 50
Chauvan, 33
Chenevard, Aimé, 52
Chereau, 41
Chicaneau, 54
China, 11-20
Chinoiseries, 36-40, 211-213
Chouard, 49
Christ College, 4
Color, and texture, 169; combination of, 151-165; complementary, 152; effect of, 144-149; effect on light, 147; effect on size of room, 146; importance of, 137-142; scheme, place of wallpaper in, 164, 165; saturation, 148, 149; values of, 161.
Conolis, 55
Cortulot, 52
Costain, 48
Couture, Thomas, 53
Couvier, 43
Crane, Walter, 110, 111, 115
Crèpy the Elder, 41
Cupid and Psyche, 51, 58, 75

Damiens, 54
Damier, 41
Dauptain, 52
Davis, R. B., 58
Defossé and Karth, 53
Delafosse, 48
Delaney, Mrs., 43
Dèlicourt, 51, 53, 54
Delft, 5
Dèsportes, 54
Dodard, 54
Domino, 3, 4, 6-9, 22
Drying of Wallpaper, 96, 97

INDEX

Dufour, 51
Dufourcoy, 25
Dugourc, 49
Dumont, 50, 53
Dupperron, Amisson, 49
Dussurguey, 50
Dye, 45, 58, 101

Eckhardt, 32
Ehrmann, 50
Embossing, 98, 99
England, 4, 5, 8-11, 21, 26, 56
Evariste, Alexandre, 51

Fan, J. J., 44
Fleeson, Peter, 68
Flock, 9, 10, 101, 102
Flowers in design, 112
Fontaine, 48
Forcoy, 25
Fortune, Robert, 18
Fragonard, 48
Fragonard fils, 51
France, 10, 21, 22
Franklin frères, 78, 79
Frèsnard, 54
Frieze, 127
Fuch, 50

Gaguet and Caffere, 54
Garnier, 41
Germany, 11, 20, 21, 43, 59
Goupy, 25
Grant Niost, 67, 68
Greenaway, Kate, 63, 64
Guèrin, 48
Guillot, J., 55
Gurat, 53

Hancock, Thomas, 67, 70
Hanging, methods of, 41, 42
Hargrave, Thomas, 67
Heeson, Plunkett, 68
Hennery, Demoiselle, 41
Hermann, 50
Hertford, Lady, 39
Holland, 6, 8, 43
Hovey, Joseph, 68
Howell, 60, 68
Huet, 44, 48
Huquier, Jacques Gabriel, 34

Jackson, John Baptist, 28-31
Jacquemart and Benard, 48
Japan, 61, 62

Laffitte, 51
Lancake, 41
Langier, 55
Langlois, 26
Lanyer, Jeremy, 10
Laurent, Joseph, 44
Lavallé-Poussin, 44
Lecomte, 34, 35, 39
Le Flaguais, 55
Le François, 10
Legendre, 54
Legrand, 49
Leroy, 52
Le Sueur, 24
Letourny, 25
Luster paper, 23

Maçon, 52
Madèr, 51, 52
Madèr fils, 51, 53
Mare, André, 87, 91
Martin, 53
Masefield, 32
Masson, 25, 54
Mathon, 41
May, William, 68
Mèry père, 44, 49
Miyer, 25

INDEX

Mognat-Perrin, 55
Molière, 52
Mongin, 50
Mongrie, 50
Monnet, Charles, 44
Montague, Lady Mary Wortley, 39
Montrille, 54
Morris, Wm., 61-64, 115, 122, 127, 139, 140
Müller, Ch-L., 53

Niodot, 41

Oberkampf, 59

Paget, 44
Pancet, 41
Panseron, Pierre, 24
Papillon, Jean, 22-26, 28, 33
Paulot and Carre, 54
Percier, 48, 51
Pèriguex, 54
Persian rugs, 122, 124
Perspective, 106, 107
Pesant, Vincent, 24
Pignet, 55
Pillement, Jean, 25
Poilly, M. N. B., 33
Polich, Martin, 52
Pompadour, Mme. de, 36, 37
Portelet, 51, 52
Potter, 60
Poyntell, William, 68
Prentis and May, 68
Prieur, 44, 48
Printing machine, 59, 60, 94-102
Prud'hon, 49

Rabier, 25
Reveillon, 43-48
Richon, 55
Ridé, 48
Riesner, 53
Robert, 47, 48
Robert, Hubert, 48
Roguié, 40
Rollers, printing, 100, 101
Rouen, 10
Roumier, 24
Rousseau Brothers, 44
Rowlandson, Thomas, 56, 57
Rugar, John, 67
Rugendas, 50

Sauvage, Joseph, 44
Savary des Breslins, 8
Scale, 125, 187-191
Schinkel, Herman, 5
Schöppler and Hartmann, 59
Semon, 52
Semone, 33
Sheraton, 56
Sietti, 44
Silhouette, 121
Sprölin, 59

Telemachus, 52
Texture, and color, 169; and size, 170
Tierce, 10

Van Loo, 48
Van Spraedonck, 44
Vaseau, 25
Vauchelet, 55
Velay, 55
Vera, 91
Vienna, 55, 59
Vincent, Thomas, 31
Vitry, 54

Wagner, 51, 53
Walpole, Horace, 29, 30, 31

INDEX

Walsh, John, 69
Washington, paper commemorating death of, 55; Cornwallis presenting sword to, 76
Washington, Martha, 77, 78
Waterloo, paper commemorating Battle of, 55
Watin, 41
Wèry, 55
Whistler, influence of, 137, 138
William, Henry, 55

Zipelius, 50
Zuber, 50-52, 58